The Eye

STRUCTURE AND FUNCTION IN DISEASE
MONOGRAPH SERIES

ABNER GOLDEN, M.D.
Series Editor

Available Volumes
Golden & Maher: THE KIDNEY
Robertson & Dinsdale: THE NERVOUS SYSTEM
Kashgarian & Burrow: THE ENDOCRINE GLANDS
Sokoloff & Bland: THE MUSCULOSKELETAL SYSTEM

STRUCTURE AND FUNCTION IN DISEASE MONOGRAPH SERIES

The Eye

Gordon K. Klintworth, M.D., Ph.D.
Professor of Pathology
Duke University

Maurice B. Landers III, M.D.
Associate Professor of Ophthalmology
Duke University

The Williams and Wilkins Company • *Baltimore 1976*

Copyright ©, 1976
The Williams & Wilkins Company
428 E. Preston Street
Baltimore, Md. 21202, U.S.A.

All rights reserved. This book is protected by copyright. No part of this book may be reproduced in any form or by any means, including photocopying, or utilized by any information storage and retrieval system without written permission from the copyright owner.

Made in the United States of America

Library of Congress Cataloging in Publication Data

Klintworth, Gordon K
   The eye.

   (Structure and function in disease monograph series)
   Includes bibliographical references.
   1. Eye—Diseases and defects. I. Landers, Maurice B., joint editor. II. Title. [DNLM:
1. Eye—Physiopathology. WW100 K65e]
RE66.K59   617.7   75-19061
ISBN 0-683-04628-4

Composed and printed at the
Waverly Press, Inc.
Mt. Royal and Guilford Aves.
Baltimore, Md. 21202, U.S.A.

FOREWORD

It is appropriate that *The Eye* by Klintworth and Landers, the fifth member of the *Structure and Function in Disease Monograph Series*, was written by authors from Duke University. A major revision of the curriculum of the School of Medicine was inaugurated with the incoming class of 1966. This permitted students to pursue subjects and courses of their own choosing to occupy fully their academic years as juniors and seniors. Medical students were thus afforded an opportunity to study many subjects in depth in a manner available at few medical schools. *The Eye* is, to a great extent, the outgrowth of a course in ophthalmic pathology that is offered by Drs. Klintworth and Landers to their medical students.

Gordon K. Klintworth graduated from the medical school of the university of the Witwatersrand in South Africa. He received postgraduate training in several disciplines in South Africa before immigrating to the United States. Dr. Klintworth came to Duke in 1962 for additional training in anatomic pathology and neuropathology. Following this, he has remained there except for a sabbatical leave in 1970 that he spent as visiting professor in the Department of Pathology of the Institute of Ophthalmology in London. While at Duke, Klintworth became particularly interested in diseases of the eye and has been responsible for Opthalmic Pathology at Duke University Medical Center since 1964.

Maurice B. Landers III, was educated at Princeton University and the University of Michigan Medical School. He did his ophthalmology training at the Jules Stein Eye Institute at the University of California in Los Angeles. He has been at Duke University since 1969. He is Associate Professor of Ophthalmology and Chief of the Retina Clinic.

Abner Golden, M.D.
Series Editor

PREFACE

The eye, like any other structure, is a unit of the entire organism prone to many diseases. The noxious agents, as well as the body's reaction to them, are fundamentally identical to those in all other tissues. Several unique characteristics of the eye, however, distinguish it from other organs. The eye is the only part of the body where the central nervous system can be viewed directly at the optic disc. It is also one of the few sites where blood vessels can be readily visualized in the living individual. The eye is one of the most exposed structures of the body and its superficial location makes it particularly vulnerable to physical and chemical injury. The crystal clear optic media permit visualization of the lesions and observation of their evolution in the living subject with greater clarity than in most tissues. Processes that interfere with the transparency of the ocular tissue impair its ability to function as a satisfactory visual organ. Some lesions which would not cause any disability in the patient if located in other parts of the body can result in severe functional disturbances, especially if located at the macula.

That an understanding of pathologic processes in any tissue or organ requires a sound knowledge of its morphologic, developmental, biochemical, and functional attributes is particularly true when one considers the eye. It contains avascular tissues like the cornea and possesses highly specialized structures that lack counterparts in other parts of the body. The unique immunologic status of the eye has permitted ophthalmologists to transplant corneal tissues successfully between antigenically different hosts with a greater degree of impunity than can be obtained in other tissues. It has also enabled investigators to study the behavior of implanted neoplastic cells in the anterior chamber.

This monograph would not have been possible without the assistance of many individuals. We are grateful to colleagues and students who offered constructive criticisms and helped with the proofreading of the manuscript. We would like to single out Drs. Jared N. Schwartz, Michael Bradbury, Frank Cashwell, Carl Fromer, J. Richard Marion III and Harold E. Shaw in this regard. Mr. Carl Bishop and Mr. Bill Boyarsky played an invaluable role with the photography. We would also like to thank Dr. Dolph O. Adams, Dr. W. Banks Anderson, Jr., and Dr. John W. Reed for providing some of the illustrative material. We would also like to express our appreciation to Mrs. Patricia Burks, Mrs. Karen

Burns, and Mrs. Frances Slocum for their patience in typing the manuscript during its various phases of development.

Gordon K. Klintworth
Maurice B. Landers, III

GENERAL REFERENCES

DAVSON, H. (Ed). The Eye, 2nd edition (6 volumes) Academic Press, New York, 1969–1974.

DUKE-ELDER, S. (Ed). System of Ophthalmology (15 volumes). C. V. Mosby, Co., St. Louis, 1958–1972.

FINE, B. S., AND YANOFF, M. Ocular Histology: A Text and Atlas. Harper & Row, Publishers, New York, 1972.

GREER, C. H. Ocular Pathology, 2nd edition. Blackwell Scientific Publications, Oxford, London, Edinburgh, Melbourne, 1972.

HOGAN, M. J., and ZIMMERMAN, L. E. (Eds). Ophthalmic Pathology: An Atlas and Textbook, 2nd edition. W. B. Saunders Co., Philadelphia, 1966.

HOGAN, M. J., ALVARADO, J. A., AND WEDDELL, J. E. Histology of the Human Eye. W. B. Saunders, Philadelphia, 1971.

MOSES, R. (Ed). Adler's Physiology of the Eye. C. V. Mosby, Co., St. Louis, 1970.

YANOFF, M., AND FINE, B. S. Ocular Pathology: A Text and Atlas. Harper & Row, Publishers, New York, 1975.

CONTENTS

Foreword v
Preface vii

1. REACTIONS OF OCULAR TISSUES TO NOXIOUS AGENTS 1
   Atrophy and metaplasia. Hypertrophy. Abnormalities in maturation. Hyperplasia, dysplasia, and neoplasia. Intracellular accumulation of material. Disorders of cell hydration (cloudy swelling, hydropic degeneration). Necrosis. Abnormalities of the extracellular constituents. Protein accumulation. Deposition of crystals in tissue. Abnormalities of the normal extracellular constituents. Deposition of metals.

2. INFLAMMATION OF THE OCULAR TISSUES 16
   Sequelae of inflammation. Special features of inflammation in different sites. Miscellaneous inflammatory diseases of unknown cause.

3. IMMUNOLOGIC DISORDERS OF THE EYE 33
   Anaphylactogenic reactions. Cell-mediated immune responses. Humoral immune responses. Autoallergies. Tissue transplantation. Immunologically related diseases.

4. VASCULAR DISORDERS OF THE EYE 44
   Blood vessels. Hemorrhage. Microaneurysms. Neovascularization. Exudates, transudates, and insudates. Hyperemia (vasodilation). Occlusovascular disease. Specific vascular diseases.

5. DEVELOPMENTAL ANOMALIES 68
   Hamartomas and choristomas. Nevi and anomalies of pigmentation. Anomalous development of parts of the eye. Syndromes.

6. NEOPLASMS AND CYSTS 89
   Neoplasms of neuroepithelium. Epithelial neoplasms. Neoplasms of muscle. Neoplasms of nerves. Lymphomas. Melanomas. Metastatic neoplasms. Miscellaneous non-neoplastic mass lesions.

7. DISEASES DUE TO MICROORGANISMS 114
   Bacterial diseases. Viral diseases. Diseases due to spirochetes.

Rickettsial diseases. Fungal diseases. Diseases due to *Chlamydia* (Bedsoniae). Parasitic diseases of the eye.

8. METABOLIC DISORDERS OF THE EYE 138
   Disorders of carbohydrate metabolism. Disorders of mucopolysaccharides. Disorders of protein and amino acid metabolism. Disturbances of lipid metabolism. Disorders of heavy metals. Miscellaneous inherited diseases with unknown metabolic defect. Oxygen toxicity. Vitamin deficiencies and excesses. Endocrine disorders. Drugs and toxins.

9. PHYSICAL AND CHEMICAL INJURIES 166
   Direct physical trauma to the eye. Surgical wounds. Wound healing and complications. Foreign bodies. Electromagnetic radiation. Thermal injuries. Ultrasonic injuries. Chemical injuries to the eye. Indirect ocular injuries.

10. ANOMALIES OF INTRAOCULAR PRESSURE 180
    Glaucoma. Hypotony (hypotonia).

11. MISCELLANEOUS CONDITIONS OF THE EYE 191
    Conjunctiva and cornea. Uvea. Retina. Lens. Vitreous body. Optic nerve. Eyelids. Orbit. Myopia.

    Index 221

1

REACTIONS OF OCULAR TISSUES TO NOXIOUS AGENTS

Tissues throughout the body, including the eye, respond in a variety of ways to noxious stimuli. Depending upon the circumstances, cells may adapt to the adverse conditions or enter a complex state in which the alterations may or may not be reversible if the injurious stimulus is removed. The chain of events which the primary pathogenic agent provokes can eventually become manifest in clinical disease.

ATROPHY AND METAPLASIA

Atrophy refers to a decrease in the size of an organ or cells which were once of mature proportion. After any severe ocular trauma, infection, or other destructive lesion, the eyeball commonly becomes atrophic and disorganized. The term phthisis bulbi has been applied to such eyes (Fig. 1.1.). The retina, choroid, and ciliary body are usually disorganized and detached from their usual location. The cornea is opaque and frequently calcified. Hyaline excrescences on Bruch's membrane (drusen), scleral thickening, and calcified cataracts are common. Metaplasia indicates transformation of one cell type into another. Under certain circumstances, bone, cartilage, striated muscle, or adipose connective tissue develops in the eye. Intraocular ossification commonly occurs in phthisic eyes many years after an original injury, but sometimes within a year. Ossification usually becomes first evident in the inner portion of the posterior choroid near the disc. It often extends into the plane between the choroid and retina. The interconnecting osseous trabeculae are usually filled with fibrous tissue, blood vessels, and degenerated remnants of choroidal stroma. Bone marrow is commonly present. It is usually a fatty marrow, but may be hematopoietic (Fig. 1.2). Adipose connective tissue may be evident within the globe in a number of ocular abnormalities such as persistent primary hyperplastic vitreous. Striated muscle and cartilage can develop within the eye in a rare neoplasm termed a teratoid medulloepithelioma. Intraocular cartilage has been reported in trisomy 13, chromosome 18 deletion, and microphthalmic eyes.

2   The Eye

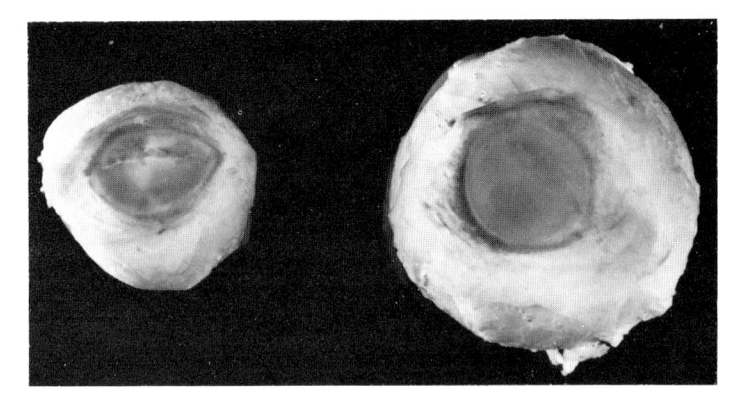

Figure 1.1. An atrophic disorganized eye (*left*) is adjacent to its normal sized fellow eye. The term phthisis bulbi is applied to such abnormal eyes.

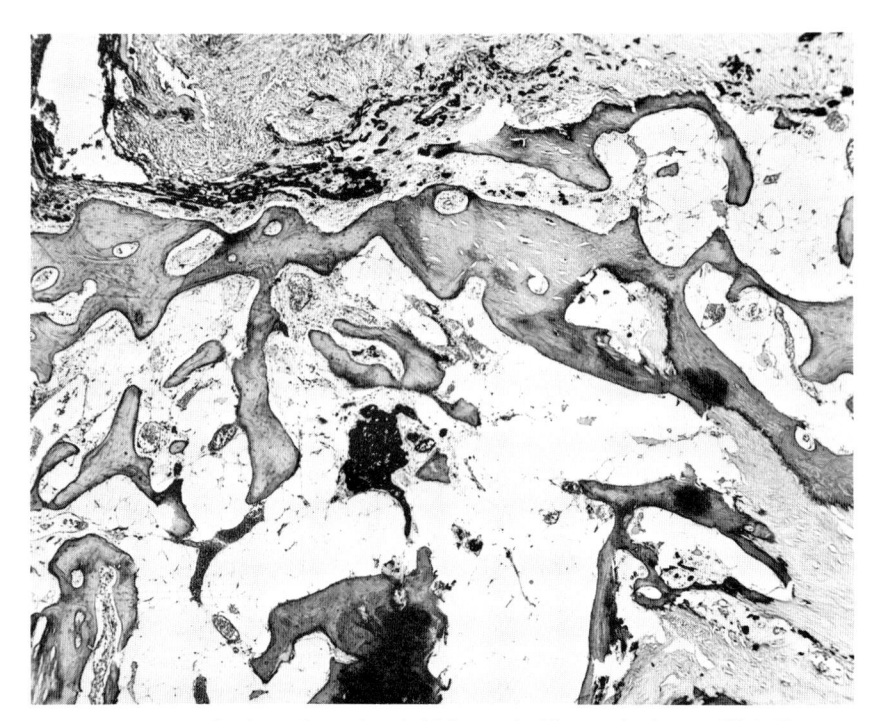

Figure 1.2. Intraocular bone from the phthisic eye in Fig. 1.1 is shown. (H & E, × 25)

HYPERTROPHY

Certain ocular cells, such as the retinal pigment epithelium, occasionally become markedly enlarged due to a numerical increase in structural and nonstructural intracellular components. Some viral diseases, especially cytomegalovirus infection, are also associated with large cells. Neoplastic cells occasionally exhibit exaggerated growth. Hypertrophied cells can result from cell fusion. Some excessively enlarged cells are multinucleated such as those associated with granulomatous inflammation.

ABNORMALITIES IN MATURATION

Under certain circumstances, the epidermis of the eyelids may develop an excessively thick keratin layer (hyperkeratosis). Usually during keratinization, the superficial epithelial cells lose their nuclei, but during imperfect keratinization, the horny layer of the epidermis may retain nuclei (parakeratosis). In areas of parakeratosis, the epidermis lacks a granular layer. Individual epidermal cells may undergo premature or atypical keratinization (dyskeratosis). This is a prominent feature in the conjunctival and corneal epithelium of an inherited disorder known as hereditary benign intraepithelial dyskeratosis that occurs in descendants of a triracial hybrid population from North Carolina.

HYPERPLASIA, DYSPLASIA, AND NEOPLASIA

In the inflammatory reaction, fibroblasts, lymphocytes, and the capillary endothelial cells undergo a numerical increase in response to increased functional demands (hyperplasia). The conjunctival epithelium may also become hyperplastic as after chronic conjunctival irritation. Hyperplasia is sometimes a consequence of infection by certain viruses as in the common wart. Hyperplasia of the skin can result in an increase in the thickness of the stratum Malpighii (acanthosis) or in an upward proliferation of papillae causing the surface of the epidermis to show irregular undulation (papillomatosis).

The term dysplasia refers to a disturbance of differentiation of an epithelial surface. It is characterized by a variation in the size, shape, and overall organization of hyperplastic cells. Such cells lack the normal cellular constituents and staining characteristics. Because the term dysplasia has come to connote a premalignant alteration, some have preferred the term atypical hyperplasia for this type of abnormality. Atypical hyperplasia can occur in the conjunctiva adjacent to the corneoscleral limbus and in the eyelid as in actinic keratosis. The term dysplasia should not be confused with dystrophy which means "ill nourished." A heterogeneous group of corneal and retinal diseases have been designated "dystrophies." Most of these are bilateral genetically determined disorders.

A neoplasm, on the other hand, may be defined as the local cellular overgrowth which results from the purposeless proliferation of intrinsically derived abnormal cells. In the eye, as in other tissues, it is sometimes difficult to distinguish between atypical hyperplasia and neoplastic proliferation, and neoplasms frequently pass through a morphologic phase of atypical hyperplasia during their development. Neoplasms have traditionally been divided into those which will not destroy their host, except by virtue of strategic placement of function. (benign) and those which are inevitably lethal to their host if unchecked by complete removal (malignant). Although such a division is important from the standpoint of diagnosis, treatment, and prognosis, the distinction between benign and malignant is not clear-cut when certain neoplasms, including some related to the eye, are considered. The biologic behavior of certain neoplasms spans a spectrum which includes benignity and malignancy.

INTRACELLULAR ACCUMULATION OF MATERIAL

Numerous substances of variable types accumulate as discrete globules, granules, or crystals within diffferent cells as a manifestation of several physiologic processes and disease states. Normally, granules are present within melanocytes, mast cells, eosinophils, and neutrophils. In pathologic states, the intracellular storage may result from defective degradation of metabolic products by cellular enzymes, such as those within lysosomes. It may also accompany a systemic metabolic disorder such as diabetes mellitus or hypercholesterolemia in which the stored material is secondary to an exposure of cells to an environment which contains excessive quantities of a metabolic precursor.

Yellow, slightly elevated, discrete cutaneous lesions consisting of numerous lipid-laden macrophages are often found in the eyelids (xanthelasmas) (Figs. 1.3, 1.4). Most occur in the upper or lower eyelid near the inner canthus. They are usually bilaterally symmetrical. Xanthelasmas often occur spontaneously, but are commonly associated with diabetes mellitus, familial hypercholesterolemia, and other disorders in which the serum lipoproteins are elevated (hyperlipoproteinemia types II and IV). They occur especially in the female in the 5th decade. Xanthelasmas are often associated with analogous xanthomas over the tendons or extensor surfaces of the limbs. An abnormal accumulation of lipid also occurs in the inherited lipidoses.

Several genetically determined diseases result in a pronounced accumulation of metabolic products within the cytoplasm of the retinal ganglion cells. The stored material imparts a degree of paleness to the retina. Because the ganglion cells are most concentrated in the region of the macula, the pallor is most pronounced in that location. The center of the macula lacks ganglion cells, thus allowing the underlying retinal

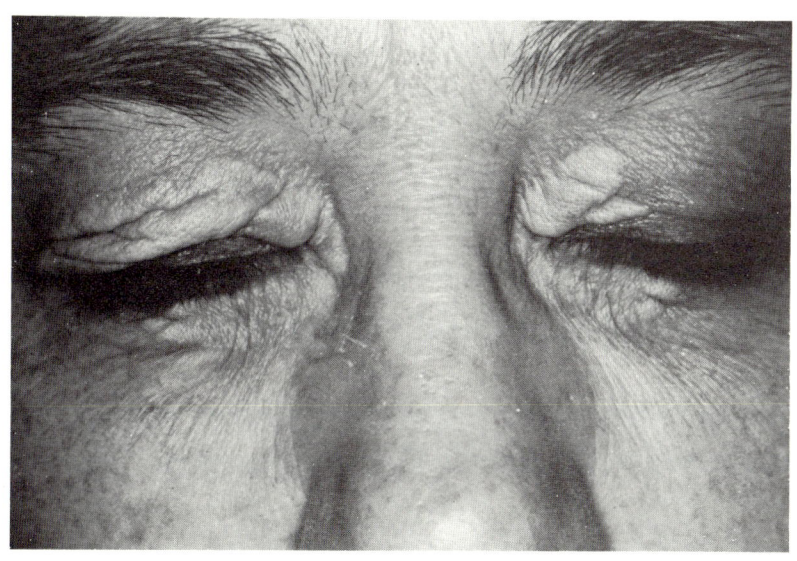

Figure 1.3. Xanthelasmas of the eyelids. Elevated lesions involve the medial aspects of the upper and lower lids of both eyes.

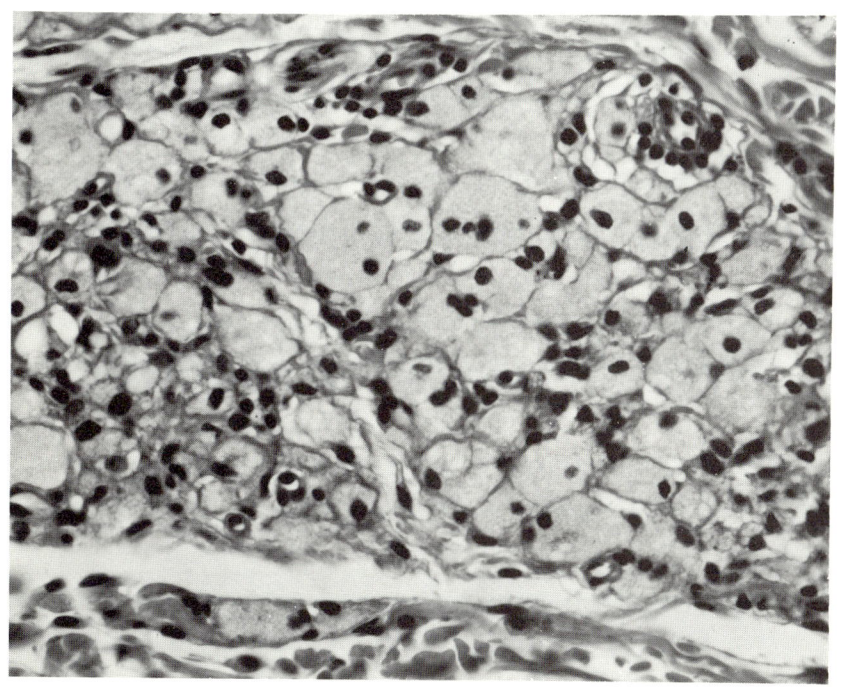

Figure 1.4. Xanthelasmas such as those in Fig. 1.3 are composed of numerous lipid-laden macrophages. (H & E, × 425)

6   *The Eye*

pigment epithelium and choroid to stand out in sharp contrast as a "cherry red spot" (Fig. 1.5).

A wide variety of intracytoplasmic inclusion bodies occur in the ocular tissues. Some cytoplasmic inclusions are related to lysosomes as in the autophagocytosis of cytoplasmic organelles. Cells, such as macrophages and those forming the retinal pigment epithelium, may become engorged with indigestible or partially digestible material phagocytosed from the environment. In many cells, lipofuscin pigment, the residual product of intracellular digestion, accumulates notably with age. Excessive amounts of lipofuscin occur in the retinal ganglia and other cells in neuronal ceroid lipofuscinosis (Batten's disease). Membranes in concentric or reticular arrangement occur normally in vacuoles within the retinal pigment epithelium. Their accumulation is induced in retinal ganglia and other cells by a variety of drugs including triparanol and chloroquine. In some inherited diseases, metabolic products, such as lipid, carbohydrate, or protein, accumulate within the cytoplasm of cells. The distribution of the stored material is often specific for the genetic disorder. Glycogen accumulates in the pigment epithelium of the iris in diabetes mellitus. Excess protein may accumulate in plasma cells (Russell bodies) in chronic inflammatory lesions. Material with staining characteristics similar to myelin sometimes accumulates in the cytoplasm of fibrous astrocytes within optic nerve gliomas (Rosenthal fibers). Observations by electron microscopy suggest that Rosenthal fibers are derived from glial filaments

Intranuclear inclusion bodies are a feature of some viral diseases that

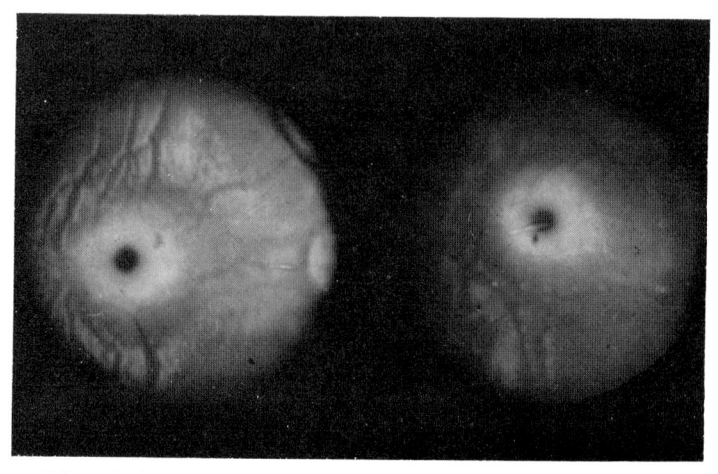

Figure 1.5. Bilateral cherry red spots of the macula. A pale white ring of abnormal retinal ganglion cells surrounds the normal central macula giving it a dark appearance.

affect the eye, including herpes zoster, herpes simplex, rubeola, and cytomegalovirus disease. Proteinaceous intranuclear inclusion bodies also occur in Waldenström's macroglobulinemia and in multiple myeloma. Other intranuclear inclusions represent cytoplasmic extensions into the nucleus.

In the dark races, the epithelium of the eyelid and conjunctiva, but not the cornea, normally contain melanin pigment at an early age. In black races, epithelial melanosis of the conjunctiva is particularly prominent around the corneoscleral limbus and usually remains stationary unlike that in "acquired conjunctival melanosis." Pigmented spots on the skin can result from exposure to sunlight, ultraviolet light, or x-rays (ephelides, freckles). The epithelium of the skin also contains excessive melanin in the café-au-lait spots of von Recklinghausen's disease of nerves. These consist of hypermelanotic macules unrelated to irradiation or pigmented nevi. Other discrete hyperpigmented macules are termed lentigines (freckle-like nevi). Two types of lentigines are recognized: juvenile and senile. The juvenile variety may be apparent at birth or develop in childhood. It may occur anywhere on the skin including the eyelid. Senile lentigines (liver spots), on the other hand, tend to occur mostly upon exposed parts of the skin, such as the forearms, dorsa of the hands, and the face, including the eyelids, in elderly people. Both types of lentigine are characterized by an increase in the pigmentation of the epidermis with elongation of the epidermal rete ridges and an increase in melanocytes. Once lentigines have appeared, they persist. They may be associated with junctional nevi and seborrheic keratosis.

DISORDERS OF CELL HYDRATION (CLOUDY SWELLING, HYDROPIC DEGENERATION)

All cells, including those in the eye, maintain an excess of intracellular potassium and a deficit of sodium with regard to the surrounding extracellular fluid. This characteristic is dependent upon energy and the sodium- and potassium-activated enzyme adenosine triphosphatase. Should this mechanism be impaired by glucose deficiency, metabolic inhibitors, or some other reason, sodium and water diffuse into the cell from the surrounding extracellular fluid and intracellular edema ensues. That this process may be reversible can be demonstrated in the isolated intact lens. An intracellular imbibition of water also accompanies the increased osmosis that follows the degradation of macromolecules into smaller particles. This phenomenon probably plays a role in the genesis of some cataracts. Intracellular edema also contributes to the retinal edema following hypoxia. The axonal and distal stumps of axons swell at the site of damage following transection.

NECROSIS

The irreversible effect of many injurious agents culminates in the death of cells or tissue within the living body. When this occurs, the cellular architecture may be preserved while cellular detail is lost (coagulation necrosis) as with infarction. The necrotic tissue may be soft and liquified (liquefaction necrosis) as in abscesses, cheesy (caseation necrosis) as in tuberculosis, or rubbery (gummatous necrosis) as in tertiary syphilis. Necrotic tissue may become the seat of putrefaction (gangrene). The necrosis may be caused by putrefying bacteria like the anaerobic clostridia of which *Clostridium perfringens* is the most common and important member. Necrosis of the exposed eyelid and conjunctiva caused by deprivation of the blood supply or by chemical action may undergo mummification due to evaporation of water, and be accompanied by a minor degree of invasion by putrefying bacteria ("dry gangrene").

After the death of cells, cellular enzymes and other constitutents are liberated into the adjacent tissue. These may destroy adjacent normal tissue components, as well as provoke an inflammatory reaction, and/or immunologic responses.

ABNORMALITIES OF THE EXTRACELLULAR CONSTITUENTS

A variety of endogenous and exogenous materials may deposit extracellularly in the ocular tissues. Some reach the tissue from distant or neighboring sites, others are produced by cells within the abnormal tissue, and yet others are derived from a breakdown of local tissue components. Factors that influence the localization of extracellular particles include physiochemical properties of the molecules such as the overall electrostatic charge, the size of the aggregates, the origin of the molecules, and the turbulence of fluid within the tissue. Certain tissue constituents act as barriers to the movement of the molecules and also contribute to their localization.

Neutral fat frequently deposits in the peripheral cornea with aging (arcus senilis) and in some of the hyperlipoproteinemias (Figs. 1.6, 1.7). The lipid is located extracellularly in the corneal stroma, and also in the **peripheral portions of Descemet's membrane and Bowman's zone. The area around the limbal blood vessels is not involved.** Mucopolysaccha-

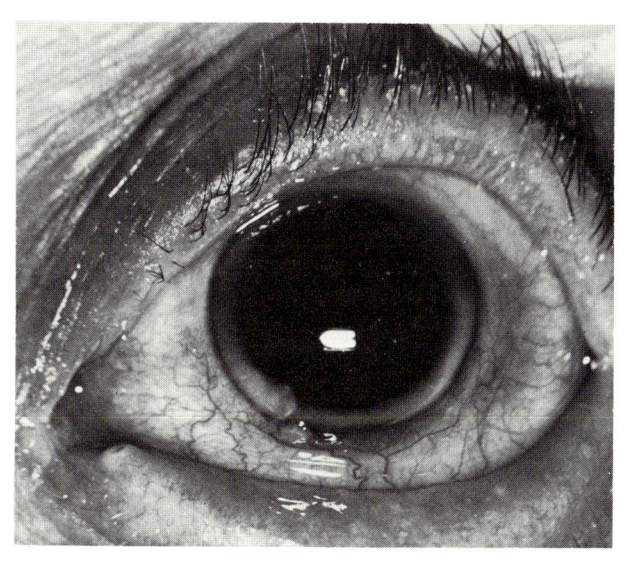

Figure 1.6. Arcus senilis. A gray ring is located in the peripheral cornea with a clear interval separating it from the corneoscleral limbus.

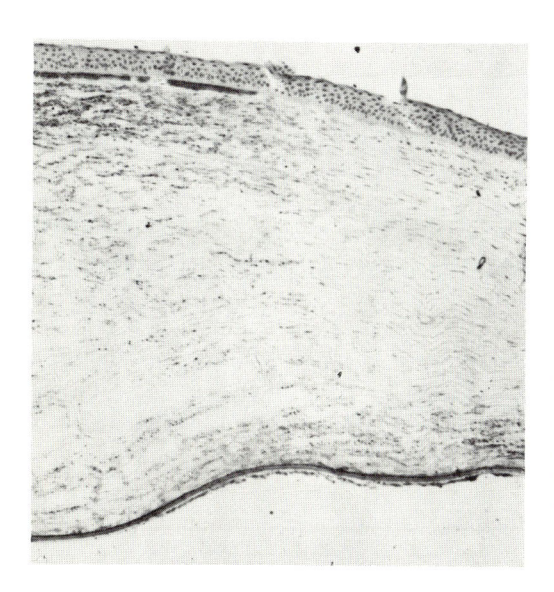

Figure 1.7. In an arcus senilis such as the one shown in Fig. 1.6, sudanophilic lipid deposits extracellularly in the corneal stroma. Note the affinity of the lipid for the periphery of Bowman's zone and Descemet's membrane. (oil red 0, × 72.4)

rides may accumulate in the optic nerve (cavernous optic atrophy, optic nerve gliomas), peripheral retina (peripheral cystoid degeneration of the retina), and the cornea (macular corneal dystrophy, systemic mucopolysaccharidoses).

PROTEIN ACCUMULATION

Fibrinoid Material

An acellular, slightly eosinophilic proteinaceous material, with staining characteristics similar to fibrin, deposits in the walls of the ocular blood vessels or connective tissue in several disorders. These include malignant hypertension, systemic lupus erythematosus, rheumatoid arthritis, rheumatic fever, polyarteritis nodosa, and the endotoxin-induced Shwartzman phenomenon. This material occurs typically in the rheumatoid nodule and in the small blood vessels in malignant hypertension and immediate hypersensitivity. Both electron microscopy and fluorescent antibody techniques indicate that fibrin is often, if not always, an important component. Fibrinoid material may represent the insudation or exudation of plasma proteins (especially fibrinogen). Fibrin accumulation also occurs if there is excessive formation or defective fibrinolysis.

Hyaline

The term hyaline has been applied on purely morphologic grounds to a translucent, homogenous, structureless eosinophilic material that occurs in the epithelium or connective tissue. Hyaline is not a single entity and at least four different types of hyaline are recognized. Such proteinaceous material deposits on the periphery of Descemet's membrane of the cornea as excrescences with aging (Hassall-Henle bodies). Identical bodies extend across the entire cornea in Fuchs' epithelial-endothelial corneal dystrophy (cornea guttata). The hyaline material in the walls of arterioles with arteriosclerosis is thought to be derived from an insudation, as well as local production. Thick bands of amorphous eosinophilic material occur in the walls of arterioles and medium sized arteries in diabetes mellitus and hypertension of long standing. The hyaline present in connective tissue in certain diseases appears to consist of excessive amounts of glycoprotein deposited between the collagen fibrils and other constituents of the tissue. Hyaline is frequently seen in connective tissue with aging. The exact mechanism of its production is not clear. Hyaline deposits that occur in the conjunctiva in chronic conjunctivitis and notably trachoma may contain amyloid.

Amyloid

Amyloid is a specific type of hyaline which is metachromatic and birefringent with an affinity for and dichroism with Congo red. It has a distinctive microfibrillary ultrastructure. It is most often extracellular.

The nature of amyloid is still uncertain. Amyloid is commonly associated with collagen but differs from it. Studies on certain types of amyloid have revealed two distinct proteins (amyloid proteins A and B). Amyloid protein A has an amino acid sequence that fails to match any described protein. The origin of the amyloid protein B has been traced to fragments of the κ light chain fragments and γ-globulin. Amyloid may deposit in the cornea in a specific genetic disease (lattice corneal dystrophy) or after certain chronic corneal diseases. It is not yet known whether it is formed locally or derived from blood-formed precursors. Amyloid occurs in the walls of blood vessels in the conjunctiva, retina, choroid, and extraocular muscles in idiopathic primary amyloidosis. Perivascular and vitreal opacities occur in one variety of inherited amyloidosis.

DEPOSITION OF CRYSTALS IN TISSUE

Crystals deposit in the ocular tissues in several diseases. In cystinosis, there is a widespread deposition of cystine crystals throughout the tissues in the body including the conjunctiva and corneal stroma. Crystals also deposit in the cornea in central crystalline corneal dystrophy (Schnyder's dystrophy) and rarely in multiple myeloma, gout, and other disorders. In gout, urate crystals may also precipitate in the lens, tarsal plates, and other ocular tissues. Calcium oxalate and tyrosine crystals sometimes deposit in cataractous lenses. After chronic retinal detachment, numerous cholesterol crystals are often present in an exudate between the sensory retina and retinal pigment epithelium (Fig. 1.8). Less commonly, oxalate crystals occur in the exudate and adjacent retina. Calcium oxalate crystals have also been observed in the retinal pigment epithelium following general anesthesia with methoxyflurane.

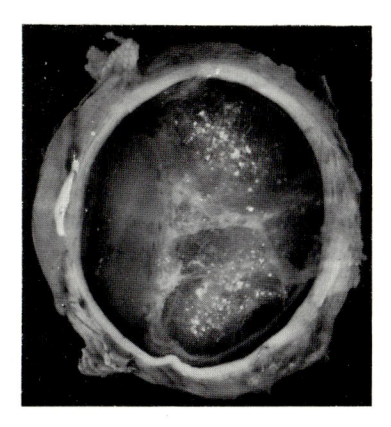

Figure 1.8. This globe with a chronic retinal detachment has numerous crystals of cholesterol between the sensory retina and retinal pigment epithelium.

ABNORMALITIES OF THE NORMAL EXTRACELLULAR CONSTITUENTS

Collagen, the major fibrous component of connective tissue, occurs in many parts of the eye. It possesses unique characteristics in certain locations. The most notable of these is the cornea where the fibrils are all of uniform diameter (360 Å) and arranged in bundles in a most orderly manner. This is in sharp contrast to the sclera which contains irregularly arranged collagen fibrils of variable diameter. The diameter of the collagen fibers may be thicker than normal, as in corneal scars. Even in pathologic states, collagen usually possesses a symmetrical (640 Å) banded pattern in the electron microscope. However, it may form fibrous or nonfibrous thread-like units with the spacing of 2560 Å (fibrous long spacing or segment long spacing collagen) due to a failure of the tropocollagen units to line up in the same direction and overlap each other by one-quarter of their length. The latter presumably occurs when the extracellular matrix is abnormal and contains excessive negatively charged polymers.

Abnormal elastic fibers occur in pseudoxanthoma elasticum, ochronosis, pingueculae, and actinic elastosis. Basophilic degeneration of the dermis (solar elastosis, actinic elastosis) occurs in man due to prolonged exposure to sunlight. It can be produced with irradiation, with ultraviolet light or x-ray. In pingueculae and actinic elastosis, variable sizes of tortuous fibers are prominent in the altered connective tissue. These fibers possess an affinity for those dyes known to stain elastic fibers, but also manifest numerous additional tinctorial attributes that distinguish them from normal elastic or collagen fibers.

DEPOSITION OF METALS

Certain metals deposit within particular ocular tissues under a variety of circumstances.

Calcium

Calcification occurs most frequently in damaged tissue in the absence of hypercalcemia (dystrophic calcification). A local tissue alkalinity plays a role in precipitating calcium. Calcification of the cornea usually involves a horizontal band of the superficial stroma (band-shaped keratopathy) (Figs. 1.9, 1.10). Calcification also occurs in necrotic areas of retinoblastomas, in drusen of the optic nerve, and in the retinal hamartomas of tuberous sclerosis. With advancing years and in certain diseases, such as Paget's disease of bone, Bruch's membrane becomes

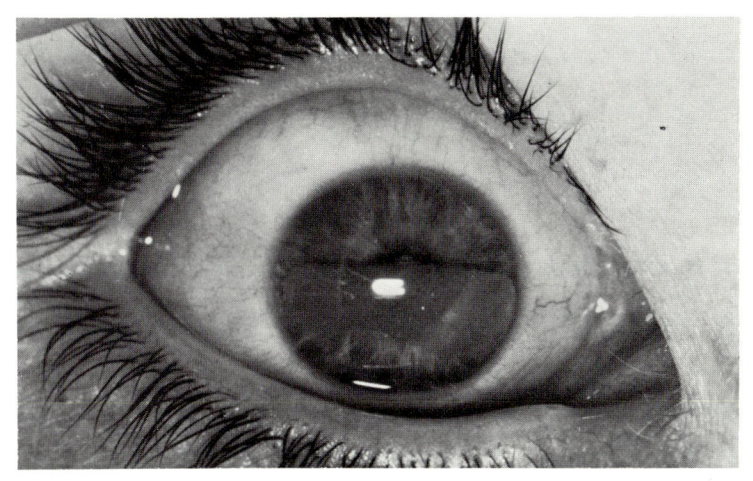

Reactions of Ocular Tissues to Noxious Agents 13

Figure 1.9. Band keratopathy. A horizontal band is visible in the cornea in the region of the palpebral fissure.

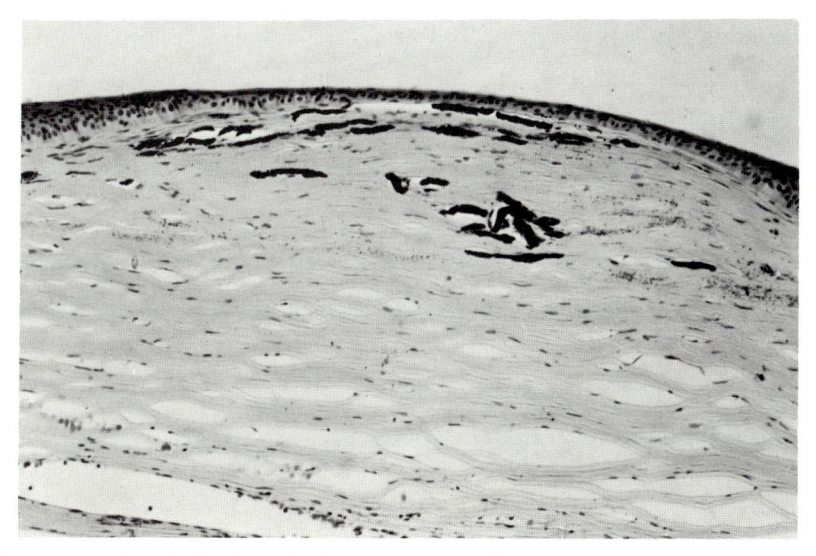

Figure 1.10. The calcified deposits that occur in band keratopathy are located extracellulary in the superficial corneal stroma. (Von Kossa, × 100)

calcified. The sclera calcifies with age and calcified scleral plaques sometimes appear immediately anterior to the medial or lateral rectus muscles in elderly patients. Cataracts frequently calcify. Calcification also occurs in the cornea and conjunctiva with hypercalcemia (metastatic calcification) as in hyperparathyroidism, increased vitamin D ingestion, and metastatic bone tumors. Because hypercalcemia may lead to cell death, it has been questioned whether calcification should rightly be divided into metastatic and dystrophic types.

Copper

A greenish brown deposition of copper occurs in the periphery of Descemet's membrane (Kayser-Fleischer ring) and in the capsule of the lens (sunflower cataract) in Wilson's disease (hepatolenticular degeneration). Copper deposits in a similar location in patients with a retained copper or bronze intraocular foreign body (chalcosis).

Iron

Particularly in individuals over the age of 50, iron deposits in the corneal epithelium, producing a horizontal line (Hudson-Stähli line). Iron also deposits in the epithelium in a ring around the conical cornea in keratoconus (Fleischer's ring) and as a pigment line ahead of a pterygium (Stocker's line) or a filtering bleb created in the treatment of glaucoma (Ferry's line).

Silver

Black granules of reduced silver deposit in the basement membrane of the corneal and conjunctival epithelium after the prolonged topical application of silver-containing eyedrops. With excessive exposure to silver in industry such as silver plating and polishing, the silver accumulates in the walls of blood vessels (argyrosis).

REFERENCES

EAGLE, R. C., AND YANOFF, M. Cholesterolosis of the anterior chamber. Albrecht von Graefes Arch. Klin. Opthalmol. 193:121–134, 1975.

FERRY, A. P. A "new" iron line of the superficial cornea. Occurrence in patients with filtering blebs. Arch. Ophthalmol. 79:142–145, 1968.

GARNER, A. Retinal oxalosis. Br. J. Ophthalmol. 58:613–619, 1974.

GARNER, A., AND TRIPATHI, R. C. Hereditary crystalline stromal dystrophy of Schnyder II. Histopathology and ultrastructure. Br. J. Ophthalmol. 56:400–408, 1972.

GASS, J. D. M. The iron lines of the superficial cornea. Arch. Ophthalmol. 71:348–358, 1964.

GOEBEL, H. H., AND FREEDMAN, A. Involvement of extraocular muscles in primary amyloidosis. Am. J. Ophthalmol. 71:1121–1127, 1971.

KLINTWORTH, G. K. Lattice corneal dystrophy: An inherited variety of amyloidosis restricted to the cornea. Am. J. Pathol. 50:371–399, 1967.

KLINTWORTH, G. K. Current concepts on the ultrastructural pathogenesis of macular and lattice corneal dystrophies. *In.* Eye, Birth Defects, Original Article Series, Vol. 7, No. 3, pp. 27-31. The National Foundation, 1971.

KLINTWORTH, G. K. Chronic actinic keratopathy: a condition associated with conjunctival elastosis (pingueculae) and typified by characteristic extracellular concretions. Am. J. Pathol. 67:327-338, 1972.

SALLMAN, L. VON, AND PATON, D. Hereditary benign intraepithelial dyskeratosis. I. Ocular manifestations. Arch. Ophthalmol. 63:421-429, 1960.

SLANSKY, H. H., AND KUWABARA, T. Intranuclear urate crystals in corneal epithelium. Arch. Ophthalmol. 80:338-344, 1968.

TSO, M. O. M., FINE, B. S., AND THORPE, H. E. Kayser-Fleischer ring and associated cataract in Wilson's disease. Am. J. Ophthalmol. 79:479-488, 1975.

WONG, V. G., AND MCFARLIN, D. E. Primary familial amyloidosis. Arch. Ophthalmol. 78:208-213, 1967.

ZIMMERMAN, L. E., FONT, R. L., AND ANDERSEN, S. R. Rhabdomyosarcomatous differentiation in malignant intraocular medulloepitheliomas. Cancer 30:817-835, 1972.

2

INFLAMMATION OF THE OCULAR TISSUES

Although inflammation of the ocular tissue is basically identical to that which occurs elsewhere in the body, certain characteristics of the eye result in specific features. The inflammatory reaction may involve the entire eye or be limited to the cornea (keratitis), conjunctiva (conjunctivitis), iris (iritis), ciliary body (cyclitis), uveal tract (uveitis), iris and ciliary body (iridocyclitis, anterior uveitis), choroid (choroiditis, posterior uveitis), retina and choroid (chorioretinitis), eyelids (blepharitis), optic disc (papillitis), optic nerve (optic neuritis, retrobulbar neuritis), lacrimal gland (dacryoadenitis), lacrimal sac (dacryocystitis), orbital connective tissue (orbital cellulitis), or intraocular structures (endophthalmitis). Endophthalmitis may be diffuse within the globe or confined to the anterior or posterior segments. When the entire globe, including the corneoscleral covering, Tenon's capsule, and adjacent orbital tissue is involved (panophthalmitis), the globe may rupture.

Inflammation may be classified according to the duration, delimitation, type of exudate, or special characteristics. Adjectives are frequently used to describe certain attributes of inflammatory lesions such as cicatrizing conjunctivitis, ligneous conjunctivitis, sclerosing keratitis, annular scleritis (involving entire corneoscleral limbus), and posterior scleritis.

The localization of the inflammatory process is of practical significance because it frequently suggests its cause. The reaction may be predominantly around one or more exciting organisms or extraneous agents. An inflammatory response that follows perforating injuries is initially at the site of entry. Certain diseases characteristically involve particular structure. For instance, in rheumatoid arthritis, episcleritis is common. The extent of the lesion depends on its cause and the severity of the inflammatory process. An inflammatory reaction may begin at one site, spread to adjacent locations, and sometimes involve the entire globe, especially if due to an infectious agent.

Several ocular structures, including the cornea, lens, and vitreous, are normally avascular and hence not capable of manifesting the hyperemia and other vascular components of the acute inflammatory response. Inflammation provoking injuries of these tissues, however, frequently excite the inflammatory response in the surrounding blood vessels. For example, corneal injuries are often accompanied by congestion of the superficial, pericorneal arterial plexus in the conjunctiva.

The inflammatory exudates contain variable amounts of protein, leukocytes, and erythrocytes and are often characteristic of the invoking agents. For example, a suppurative reaction may be caused by pyogenic microorganisms, copper, silver nitrate, and other substances which provoke a marked neutrophilic infiltration and necrosis of tissue. When exudates and inflammatory cells accumulate in the ocular tissue, an indication of their presence may be evident clinically. Inflammatory cells and a protein-rich exudate in ocular tissue that is normally transparent cause a variable degree of cloudiness. Leukocytes suspended in the aqueous humor may produce a white-yellow sediment in the anterior chamber (hypopyon) (Fig. 2.1). Although this commonly accompanies a corneal or intraocular pyogenic infection, its presence does not necessarily indicate intraocular infection. A sterile hypopyon may accompany an infected corneal ulcer ("hypopyon ulcer"). The aqueous humor normally contains the soluble constituents of blood but very little protein. Excess protein in the aqueous humor, as in iridocyclitis, results in a translucence of the aqueous that can be seen with the biomicroscope (slit lamp). This is referred to as an aqueous flare (Tyndall phenomenon). If much fibrin is

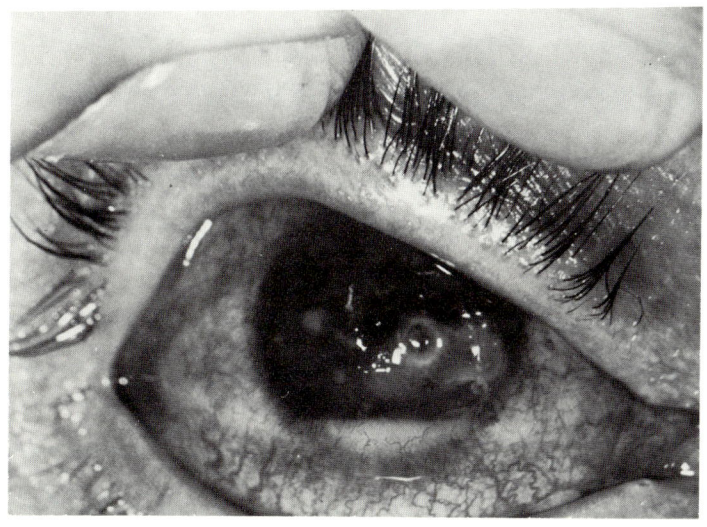

Figure 2.1. Hypopyon. A corneal ulcer has produced a layering out of leukocytes in the inferior aspect of the anterior chamber.

present in the exudate in the anterior chamber, it may clot within the eye or after aspiration. Inflammatory cells, particularly monocytes and macrophages, tend to agglutinate and may deposit in the eye where they can be visualized clinically. A variety of terms have been applied by ophthalmologists to the aggregates of inflammatory cells that are inherent to different ocular structures: aggregates on the iris margin (Koeppe's nodules), on the iris surface (floccules of Busacca, Busacca nodules), on the inner surface of the retina (retinal or preretinal precipitates); and on the corneal endothelium (keratic precipitates) (Fig. 2.2). Different types of keratic precipitates occur. Agglutinated aggregations of mononuclear cells and macrophages adhere to the corneal endothelium in large deposits that clinically resemble fat (lardaceous, waxy, or mutton fat keratic precipitates). When numerous, such deposits may form a continuous layer (conglomerate keratic precipitates). When composed of lymphocytes and plasma cells, they appear as small white punctate accumulations. Large deposits are more characteristic of granulomatous inflammation. Although agglutination is not a feature of

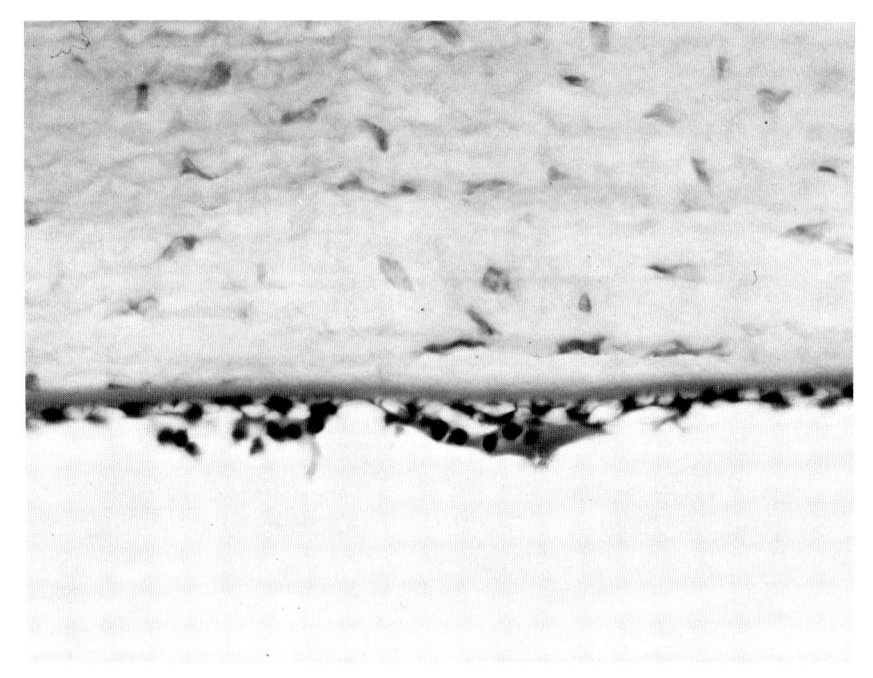

Figure 2.2. Leukocytes, and particularly mononuclear cells, frequently adhere to the corneal endothelium in iritis. When this occurs, the deposits may be detected clinically with the slit lamp. (H & E, × 680).

neutrophils, the adherence of small groups of these cells to the cornea may result in a frosted appearance.

After the exudative phase of the inflammatory response, a proliferative one begins in which there is regeneration of destroyed tissues, new vessel formation, and repair of the defect. The repair of damaged tissue depends upon several factors. These include the nature of both the noxious agent and the injured tissue, and the quantity of tissue damage, as well as the persistence of noxious agents such as microorganisms or foreign bodies.

Phagocytosis of necrotic debris, particularly by large mononuclear cells, occurs. The ingested material is digested by the lysosomal system of the macrophages with indigestible material remaining as residual bodies.

The ability of the destroyed structures to regenerate depends upon the limitations of the cells and the architecture. Some ocular cells such as the lens epithelium, corneal epithelium, and endothelium can regenerate and grow over a denuded surface. The corneal endothelium may grow backwards over the anterior surface of the iris and deposit Descemet's membrane where it is not normally located. The retina is not capable of regeneration and repair results in a mixed glial and collagenous scar. Near the ora serrata, the epithelia of the ciliary body and retinal pigment epithelium commonly proliferate, especially when the ciliary body is pulled inward by a cyclitic membrane. Excrescences ("drusen") may appear beneath the retinal pigment epithelium on Bruch's membrane. An excessive proliferation of epithelium may resemble a neoplasm (pseudoepitheliomatous hyperplasia).

Proliferating fibroblasts may occasionally be pleomorphic and accompanied by much extracellular mucopolysaccharide and many inflammatory cells. This reaction, termed nodular fasciitis (pseudosarcomatous fasciitis), may involve the orbit and rarely the conjunctiva.

The granular or cobbled profile which the complex of newly formed capillaries imparts has resulted in its designation as granulation tissue. A variable amount of granulation tissue forms in different wounds, being minimal in noninfected surgical wounds with closely approximated wound edges, but pronounced in ragged wounds or ulcers. The amount of granulation tissue formed may be pronounced and result in a distinct mass containing scattered inflammatory cells ("granuloma pyogenicum"). This reaction is frequently observed in the eyelids and conjunctiva.

The neovascularization of inflammation not only occurs in the inflamed area but in adjacent tissues. Newly formed vessels in the iris may enter the angle of the anterior chamber, while those in the ciliary body and adjacent structures may enter the vitreous, and those in the optic nerve head and adjacent retina may extend into the posterior vitreous. Choroidal blood vessels may extend into subretinal exudates when a break in Bruch's membrane exists.

Eventually, most blood vessels obliterate and disappear. The amount of extracellular collagen accumulates and the fibroblasts become less prominent by reverting to a quiescent state (fibrocytes) or dying. Later, the collagenous tissue merges into a structureless glassy mass (hyalinization). The fibrosis forms scars in which collagen has a thicker diameter than normal even in the cornea.

A response to slow acting, persistent, injurious agents is commonly accompanied by evidence of repair. Exudation is less in chronic inflammation than in acute inflammation. The cellular infiltrate consists especially of mononuclear cells, lymphocytes, plasma cells, and histiocytes. Granulomatous inflammation has come to have a limited connotation that implies inflammation dominated by macrophages and their derivatives such as epithelioid and giant cells. It is notably associated with cell-mediated hypersensitivity. In the eye, it appears with sympathetic ophthalmia, tuberculosis, sarcoidosis, leprosy, toxoplasmosis, syphilis, and certain fungal and other diseases. When dermoid or epidermoid cysts rupture, the extruded contents provoke a granulomatous reaction with foreign body giant cells and prominent histiocytes.

SEQUELAE OF INFLAMMATION

The inflammatory reaction may have a deleterious effect upon the maintenance of the eye as a useful visual organ. Structures such as the cornea, retina, trabecular meshwork, and lens may be damaged. Tissue may be destroyed by the injurious agent, by the responding inflammatory cells, and by the enzymes liberated by injured cells. A fibrovascular sheet commonly develops along the anterior surface of the iris (rubeosis iridis). It may extend into the angle of the anterior chamber, as well as across the pupil (pupillary membrane). Contraction of fibrous tissue along the anterior surface of the iris may evert the iris outward so that the pigment epithelium extends beyond the pupillary margin (ectropion uveae) (Fig. 2.3), while contracted fibrous tissue on the inner side may invert the pupillary margin (entropion uveae). Adhesions commonly occur between different structures such as the iris and cornea (peripheral anterior synechiae), the iris and lens (posterior synechiae), and the retina and choroid. Scars may obliterate the anterior chamber as well as incarcerate the iris and lens. Fibrous tissue may extend between different ocular tissues and appear clinically as membranes (e.g. retrocorneal, preretinal, retrovitreal, iridovitreal, and vitreocorneal). Excessive fibro-

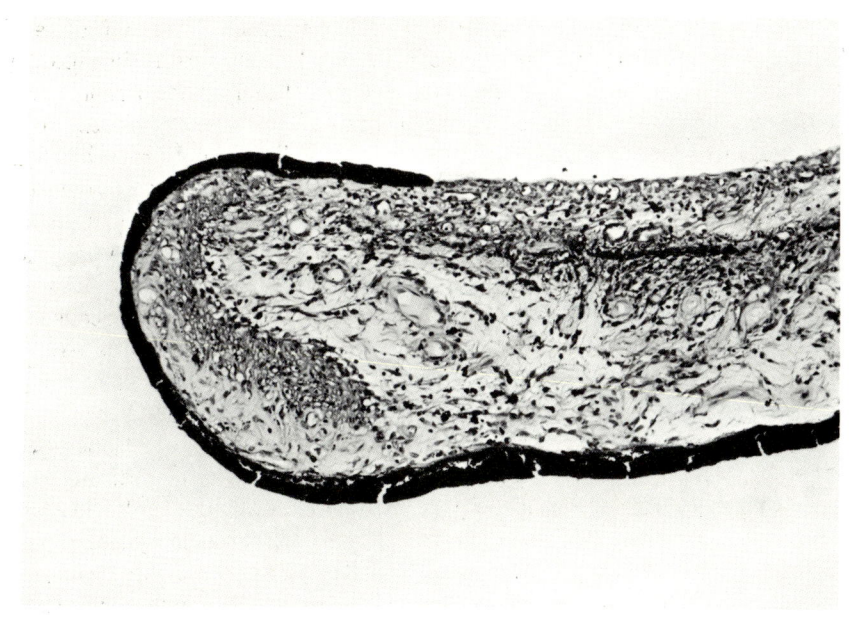

Figure 2.3. Ectropion uveae. The pigment epithelium has extended around the pupillary margin of the iris as a result of contraction of fibrovascular tissue on the anterior surface of the iris. (H & E, × 100).

sis may result in keloids of the cornea, conjunctiva, or eyelids. Adhesions in the filtration angle or pupil may impede flow of aqueous humor and result in glaucoma. Contraction of a transocular diaphragm (cyclitic membrane) may detach the ciliary body from the sclera (except at the scleral spur) and the sensory retina from the retinal pigment epithelium. Fibrosis and disorganization of the ciliary body may obliterate blood vessels and result in an impaired formation of the aqueous humor and hypotony. The integrity of damaged tissues may be weakened to such an extent that they are unable to withstand the normal or exaggerated physiologic stresses and strains.

SPECIAL FEATURES OF INFLAMMATION IN DIFFERENT SITES

Conjunctiva

The conjunctiva and cornea are particularly prone to exogenous irritants like dust and gases. Congestion of the conjunctival blood vessels, excessive tearing, and a variable conjunctival exudate occur with acute conjunctivitis, the most common disease of the eye. The cells within the exudate vary with the cause. The exudates in acute inflammation accumulate in the conjunctival sac and may coagulate and form a membrane on the surface of the eye. In an acute inflammatory response to a potent necrotizing toxin, such as the diphtheria exotoxin, a pseudomembrane composed of precipitated fibrin, necrotic epithelium, and the inflammatory cell infiltrate develops. This pseudomembranous type of reaction occurs not only with diphtheria, but with staphylococcal conjunctivitis, epidemic keratoconjunctivitis, chemical burns, and the Stevens-Johnson syndrome. A purulent conjunctival exudate may accompany an acute bacterial conjunctivitis. It is a feature of certain conjunctivitides that occur in the neonatal period due to gonococcus and other organisms (ophthalmia neonatorum). Viral conjunctivitis usually produces a minimal exudate, while a scant ropy discharge occurs with allergic conjunctivitis.

Cultures and cytologic examinations of conjunctival secretions are of great diagnostic aid. Eosinophils are abundant in atopic conjunctivitis and especially vernal conjunctivitis. Polymorphonuclear leukocytes are prominent in many acute infections due to bacteria or chlamydia. A preponderance of lymphocytes are characteristic of viral infections and chronic inflammation. Multinucleated epithelial cells with or without intracellular inclusion bodies are a feature of some viral conjunctivitides.

Lymphocytic and plasma cell infiltrates of the conjunctiva and hyperplasia of the conjunctival lymphoid tissue are common in certain pathologic states. The follicles may appear clinically as avascular nodules (follicular reaction). Some chlamydia (e.g. trachoma and inclusion conjunctivitis), viruses (e.g. adenoviruses and Newcastle disease of fowls), and chemicals (e.g. eserine and atropine), provoke the proliferation of lymphocytes including the formation of subepithelial lymphoid follicles with germinal centers in the conjunctiva. The term follicular conjunctivitis is applied to this reaction. Hyperplasia of the lymphoid tissue normally present in the plica semilunaris causes a diffuse swelling of that structure. Lymphoid tissue is normally inconspicuous in the human conjunctiva until the first few months of life. A follicular conjunctivitis does not occur in the neonate in response to stimuli that typically provoke it in adults.

The conjunctival epithelium is normally anchored at intervals to the underlying connective tissue, and subepithelial accumulations of exudate, edema, or vascular engorgement displace the epithelium between the sites of fixation forming prominent rounded or flattened papillae. This is a feature of certain types of conjunctivitis (papillary conjunctivitis). It occurs in vernal conjunctivitis, polycythemia, and acute bacterial and allergic conjunctivitis (Fig. 2.4). In neonates, as well as in the aged,

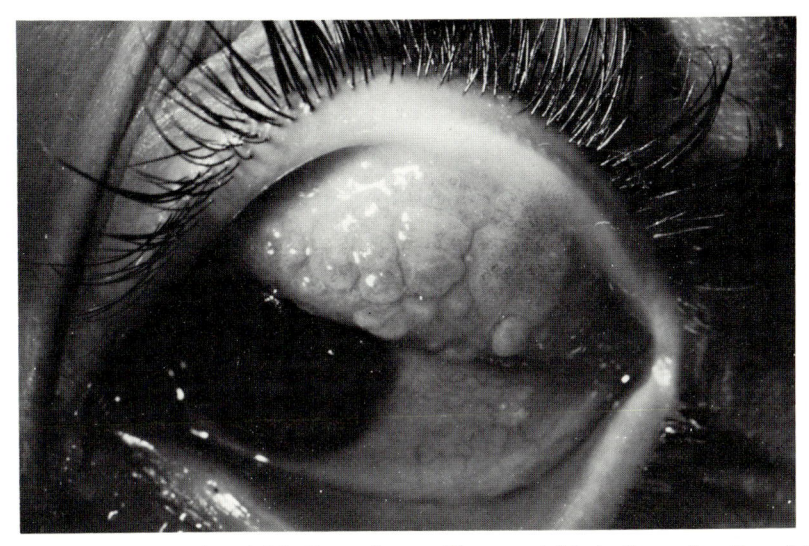

Figure 2.4. Vernal conjunctivitis. Large flat papillae are visible in the conjunctiva of the upper lid.

stimuli that cause a follicular conjunctivitis often provoke a papillary conjunctivitis instead. A rare variety of chronic recurrent conjunctivitis in young children is characterized by the formation of much hyalinized fibrous tissue (ligneous conjunctivitis).

When the conjunctival epithelium is lost in conditions such as trachoma, caustic burns, Stevens-Johnson syndrome, and benign mucous membrane pemphigoid, adhesions develop between the apposed palpebral and bulbar conjunctiva (symblepharon). The conjunctival cul-de-sac may become obliterated and alter the normal flow of tears. The cornea commonly remains exposed.

Cornea

Inflammation of the cornea may affect primarily the corneal epithelium and adjacent stroma (superficial keratitis) or be limited mainly to the corneal stroma (interstitial keratitis). Interstitial keratitis occurs in a variety of conditions including congenital syphilis, tuberculosis, and leprosy. Deep keratitis is usually associated with an intense lymphocytic and vascular invasion into the substantia propria from the limbus. It is thought to be an immunologic reaction. With acute inflammation of the cornea, a cellular infiltrate occurs with loss of corneal transparency, pain, and ciliary congestion. Blood vessels may invade the normally avascular cornea. With superficial keratitis, ulceration may occur. The loss of tissue after a corneal ulcer may cause the cornea to bulge forward (kerectasia, ectatic scar). resulting in refractive errors. The cornea may also become weakened to such an extent that Descemet's membrane bulges externally into the ulcer (descemetocele). Corneal perforation sometimes occurs and may become closed by the adherence or prolapse of iris into the defect. A corneal ectasia may be lined by uveal tissue (corneal staphyloma). Adhesions may develop between the marginal corneal ulcer and a fold of conjunctiva (pseudopterygium). Corneal scars interfere with the transparency of the cornea. Inflammation of the cornea is usually accompanied by iritis and often a hypopyon. The peripheral cornea may ulcerate ("ring ulcer"), especially in Wegener's granulomatosis, periarteritis nodosa, and rheumatoid arthritis. Such lesions may be due to vascular obliteration of the anterior ciliary arteries.

A slowly progressive chronic corneal ulcer with undermined edges (chronic serpiginous ulcer, Mooren's ulcer) begins near the corneoscleral limbus, especially in the elderly. It is associated with a prominent mononuclear leukocytic infiltrate and has a tendency to extend across the cornea. No pathogens have been consistently isolated and the condition is thought to be due to stenosis or occlusion of the pericorneal blood vessels.

Sclera

With acute inflammation of the episclera between the insertion of the rectus muscles and the corneoscleral limbus, the hyperemic vessels are clinically evident (Fig. 2.5). When the anterior portion of the sclera is inflamed, the reaction frequently extends into the cornea.

Uvea

In certain diseases, the uvea is preferentially involved. An aqueous flare, keratic precipitates on the cornea, and edema of the iris are features of an acute anterior uveitis. In posterior uveitis, vitreous opacities occur and the retina overlying the lesion may become involved and even detach due to an associated exudate. Choroiditis may be diffuse or focal. Multiple lesions are usually of hematogenous origin. Several systemic diseases are associated with a nongranulomatous uveitis. These include: ankylosing spondylitis (Marie-Strümpell's disease), Still's disease, Behçet's disease, prostatitis, ulcerative colitis, regional enteritis, and rheumatoid arthritis. An anterior uveitis may be due to herpes simplex, herpes zoster, and influenza. However, in many instances, there is no obvious evidence of the causative agent of uveitis. A granulomatous uveitis occurs in brucellosis, tuberculosis, sarcoidosis, toxoplasmosis, sympathetic ophthalmia, and Vogt-Koyanagi-Harada syndrome. In

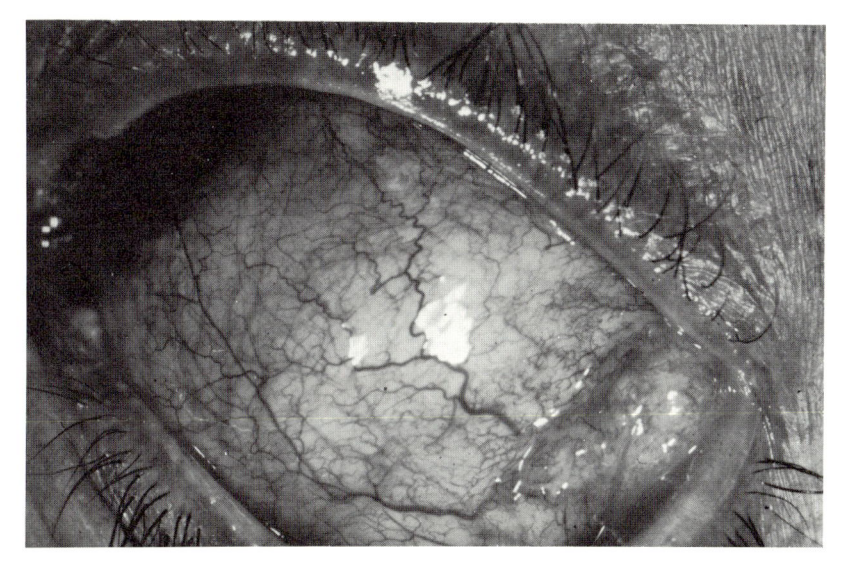

Figure 2.5. Episcleritis with hyperemic vessels visible in and beneath the conjunctiva.

granulomatous uveitis, nodules containing reactive retinal pigment epithelium, epithelioid cells, and macrophages may develop beneath Bruch's membrane and the retinal pigment epithelium (Dalen-Fuchs' nodules).

Uveitis is a frequent component of Reiter's syndrome. An idiopathic form of uveitis associated with differences in the color of the iris in the two eyes is known as heterochromic uveitis. It is usually unilateral and painless. The color change due to loss of pigment is rarely evident at birth. The condition is not associated with an acute red eye. Chronic nongranulomatous iridocyclitis with few lymphocytes and plasma cells occurs in the lighter colored eye. This condition is associated with atrophy and depigmentation of the iris and ciliary body. It is often complicated by a posterior subcapsular cataract and secondary glaucoma. Because chronic iridocyclitis causes atrophy and depigmentation of the iris, the condition may not be a specific entity but rather a manifestation of recurrent uveitis. A high percentage of patients with ankylosing spondylitis eventually develop a recurrent mild, acute uveitis which may be the presenting clinical manifestation of the condition. Anterior uveitis and episcleritis occur in many cases of ulcerative colitis.

Retina

The sensory retina commonly becomes displaced from its normal location by an inflammatory exudate or by contraction of intraocular fibrous tissue. In the retina, scarring results from the proliferation of fibrous astrocytes (Müller's cells) which may give rise to diffuse intraretinal, preretinal, or postretinal gliosis. Occasionally, the proliferation may be so extensive as to resemble a neoplasm (massive pseudogliomatous proliferation, "pseudotumor of the retina"). The retinal pigment epithelium proliferates particularly at the margins of certain types of chorioretinitis.

Lens

In certain intraocular inflammatory lesions, cataract formation is common. An inflammatory cell infiltrate does not extend into the lens if the capsule is intact, but may do so if the lens capsule is damaged as a result of trauma or infection.

Vitreous

Leukocytes infiltrating the vitreous commonly follow the planes of the filaments within it. A vitreous abscess (Fig. 2.6) rarely disperses but organizes into fibrous tissue often resulting in detached retina.

Eyelids

Acute inflammation of the Meibomian gland (internal hordeolum) and acute folliculitis of the Zeis gland (external hordeolum) are common. Chronic granulomatous inflammation with Langhans' or foreign body giant cells centered around the Meibomian or Zeis glands commonly occur in the eyelid (chalazion). Some are preceded by an acute

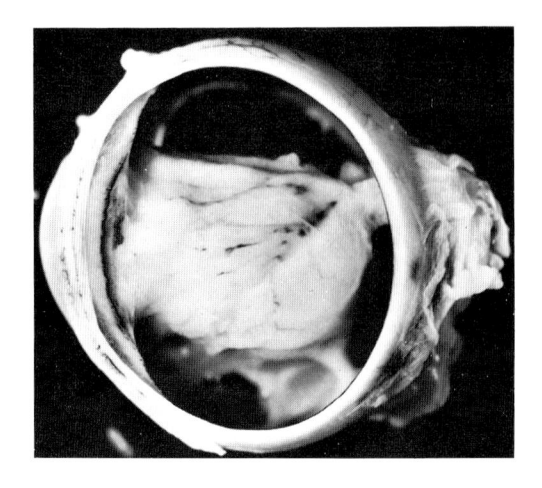

Figure 2.6. Vitreal abscess. The vitreous of this eye is filled with a purulent exudate.

inflammatory reaction. A chalazion produces a localized, painless swelling of the lid which may press on the cornea and distort vision (Figs. 2.7, 2.8).

Chronic inflammation of the eyelids often begins in childhood. Superficial scaling of the skin of the lid margins is most often due to seborrhea. The epithelium of the eyelid may ulcerate and result in a

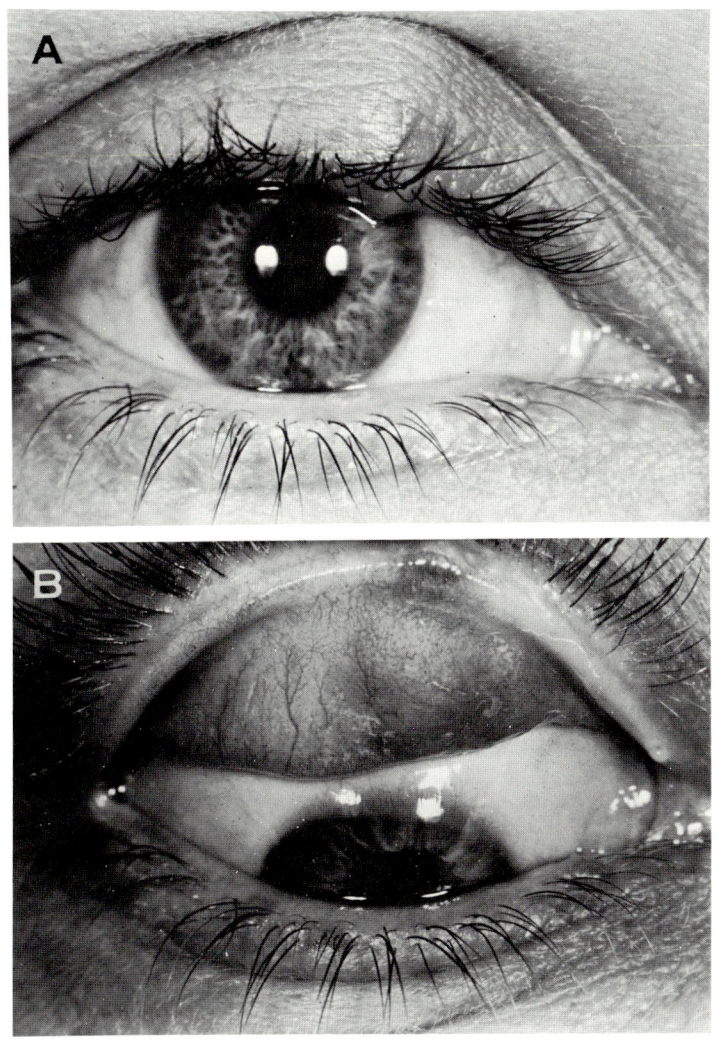

Figure 2.7. Chalazion. A localized swelling is present in the upper lid (*A*). Eversion of the upper lid shows the involvement of the lid margin and adjacent conjunctiva (*B*).

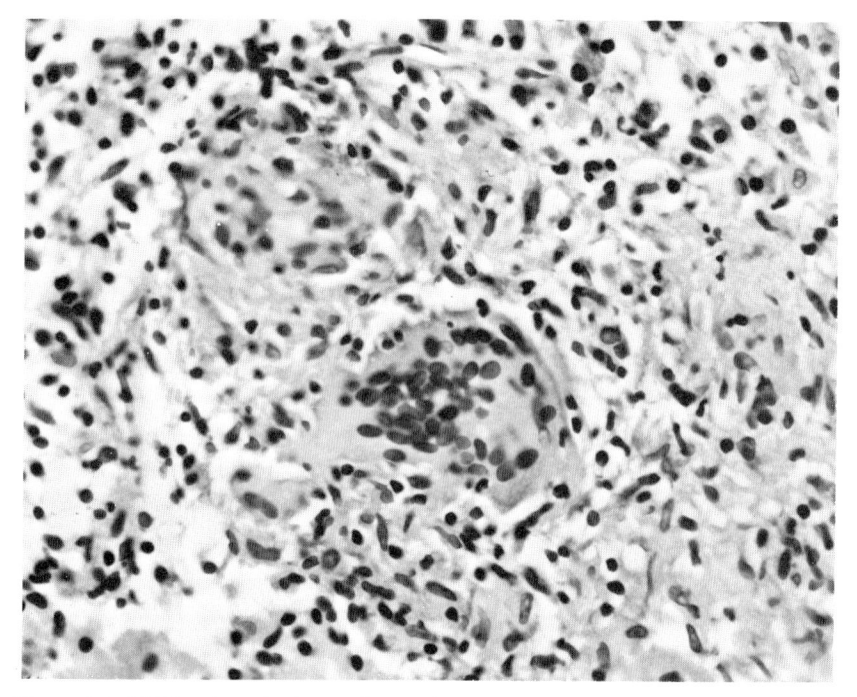

Figure 2.8. Multinucleated giant cells, such as the one shown here, are common in chalazia. (H & E, × 400).

patchy loss of lashes. The condition may be aggravated by secondary infection, especially by staphylococcus. Conjunctivitis is usually associated with this condition, and occasionally keratitis occurs.

Scars of the eyelids, as occur with infections, burns, and lacerations, may distort them, thus causing ectropion and exposure of the bulbar and papebral conjunctiva, eventually resulting in exposure keratopathy. Inward turning of the eyelids (cicatrical entropion) directs the eyelashes towards the globe (trichiasis) which may irritate the conjunctiva and frequently predispose it to secondary infection. This most often occurs after irradiation, trachoma, erythema multiforme, chemical burns, and lacerations.

Orbit

The term "inflammatory pseudotumor of the orbit" although unsatisfactory, has a clinical connotation. There is no universally acceptable definition of the term. It indicates an orbital lesion which mimics a neoplasm clinically but which histopathologically represents an idiopathic chronic inflammatory reaction associated with a variable degree of fibrosis. It is a common cause of exophthalmos and partial immobility of the eyeball. Edema of the lids and conjunctiva frequently occur. Any part of the orbit may be involved. The condition is mostly unilateral, but may occasionally become bilateral after a delay of several months to many years. The components of the lesion vary from case to case. A mixed inflammatory cell infiltrate is common but some cellular elements, such as lymphocytes or plasma cells may predominate. Such orbital inflammatory lesions may herald the onset of Wegener's granulomatosis.

Inflammation of the lacrimal gland is usually nonsuppurative, and one of the sequellae is diminished tear formation. The lacrimal gland is frequently involved in sarcoidosis.

MISCELLANEOUS INFLAMMATORY DISEASES OF UNKNOWN CAUSE

Reiter's Syndrome (Fliessinger-Leroy Syndrome)

The combination of recurrent acute conjunctivitis and/or iridocyclitis and urethritis and nonsuppurative polyarthritis in young adults, especially males, is termed Reiter's syndrome. The subacute or chronic arthritis mainly involves the sacroiliac joints and those of the lower limbs. Involvement of the feet commonly results in plantar fasciitis and bony plantar spurs. Iritis may lead to glaucoma. Lesions of the skin and mucous membrane are common and heal without scars. Pleurisy, cystitis, pericarditis, retrobulbar neuritis, and cardiac valvular lesions occasionally occur. Epidemiologic evidence suggest that Reiter's syndrome is an infectious venereal disease, but some cases may follow bacillary dysentery.

Behçet's Syndrome

A syndrome characterized by chronic relapsing oral and genital ulceration was described by Behçet in 1937. Since the original description, manifestations in other tissues have become recognized. The lesions are often painful and heal with scarring. Males are affected more commonly than females. Most manifestations are consistent with vascular occlusions secondary to a necrotizing angiitis or thrombosis. The ocular lesions are variable and iritis with hypopyon is common. A high percentage of cases become blind and fatalities may occur.

Cat Scratch Fever

In this disease, a localized pustule develops at the site of a cat scratch or bite. This is followed by a regional suppurative lymphadenitis. The eyelid and/or conjunctiva and preauricular lymph nodes may be involved (Parinaud's syndrome).

Sarcoidosis

Sarcoidosis is a disease predominantly of young adults. It occurs more frequently in blacks than whites in North America. This idiopathic systemic noncaseating granulomatous disease may involve any organ of the body including the ocular and orbital tissues. Sarcoidosis has a predilection for the anterior segment of the eye, where discrete granulomas occur in the anterior uvea. Chronic uveitis is common. Vitreous hemorrhage resulting from vitreal neovascularization occurs. The lids, conjunctiva, lacrimal glands, retina, and optic nerves may also be involved. When the eye or lacrimal gland is affected, the condition is commonly bilateral. Hypercalcemia and band keratopathy are common. Sarcoidosis is a cause of bilateral enlargement of the lacrimal and salivary glands (Mikulicz's syndrome).

Wegener's Granulomatosis

Wegener's granulomatosis is characterized by necrotizing granulomas which involve both arteries and veins often leading to thrombosis. The lesions may occasionally contain giant cells. Original reports noted that this condition involved the respiratory tract and kidneys. Lesions may affect the orbit and cause proptosis. Coagulative necrosis of the optic nerve and optic atrophy may ensue causing decreased visual acuity or blindness. Ptosis, edema of the lid, and conjunctiva, as well as ulceration of the eyelids and cornea, may also occur. Orbital involvement may precede systemic manifestations. An extremely progressive variety of the condition may be widespread with involvement of the upper respiratory tract, kidneys, and elsewhere, or the disease may be limited to one site such as the lung, orbit, or nasopharynx (midline lethal granuloma).

Histiocytosis X

The term "histiocytosis X" includes a group of uncommon conditions associated with diffuse histiocytic hyperplasia. These were previously designated eosinophilic granuloma, Hand-Schüller-Christian disease, or Letterer-Siwe disease, but are now thought to represent different stages in the spectrum of the same basic disease process. All variants contain abundant stellate shaped Langerhans cells with characteristic cytoplasmic organelles ("Birbeck granules"). These cells differ from, and should not be confused with, the Langerhans giant cell. The orbital bone may be involved in all varieties.

A relatively benign form which affects children and young adults is usually localized to a single bone and occasionally an orbital bone (eosinophilic granuloma). The osteolytic lesion is evident on radiologic examination. Eosinophils and histiocytes predominate in the usually soft and friable lesions. Lipid-laden histiocytes are not a feature.

A more severe form (Hand-Schüller-Christian disease) usually affects individuals during childhood and typically presents with the triad of exophthalmos, bony defects in the calvarium, and diabetes insipidus. Inveolvement of the tuber cinereum and hypothalamus results in diabetes insipidus. Involvement of the orbit or its surrounding bone causes unilateral or bilateral exophthalmos and sometimes ophthalmoplegia and papilledema. Multiple confluent osteolytic skull defects occur especially in the temporal parietal bone. The disease is characterized by accumulations of numerous lipid-laden histiocytes in the various organs. These contain a high content of cholesterol, cholesterol esters, and neutral fat. A variable amount of fibrosis and inflammatory cell infiltrate is associated.

A usually fatal form of histiocytosis X (Letterer-Siwe disease) affects infants and very young children. It is characterized by fever, osseous lesions, and a widespread infiltration of non-lipid-containing histiocytes. Hemorrhage and hepatosplenomegaly also occur. It has a rapidly progressive course.

Nevoxanthoendothelioma (Juvenile Xanthogranuloma)

This condition usually manifests mainly as a benign dermatologic disease in the first 5 years of life. Cutaneous papules and nodules of variable size occur, especially on the head and neck. Although the lesions may be restricted to the skin, other tissues may be involved, particularly the anterior uveal tract. Involvement of the iris results in heterochromia iridis, glaucoma, and recurrent hyphema. The orbit may also be affected. The granulomatous lesions are characterized by abundant histiocytes and prominent multinucleated giant cells (Touton giant cells) both of which contain lipid.

REFERENCES

BLODI, F. C., AND GASS, J. D. M. Inflammatory pseudotumor of the orbit. Trans. Am. Acad. Ophthalmol. Otolaryngol. 71:303–323, 1967.

DAWSON, C. R., SCHACHTER, J., OSTLER, H. B., GILBERT, R. M., SMITH, D. E., AND ENGLEMAN, E. P. Inclusion conjunctivitis and Reiter's syndrome in a married couple. Arch. Ophthalmol. 83:300–306, 1970.

FERRY, A. P., AND LEOPOLD, T. H. Marginal (ring) corneal ulcer as presenting manifestation of Wegener's granuloma — a clinicopathologic study. Trans. Am. Acad. Ophthalmol. Otolaryngol. 74:1276–1282, 1970.

FONT, R. L., AND ZIMMERMAN, L. E. Nodular fasciitis of the eye and adnexa: a report of ten cases. Arch. Ophthalmol. 75:475–481, 1966.

FRAYER, W. C. Reactivity of the retinal pigment epithelium. Trans. Am. Ophthalmol. Soc. 64:586–643, 1966.

GARNER, A. Pathology of "pseudotumors" of the orbit: a review. J. Clin. Pathol. 26:639–648, 1973.

PERKINS, E. S. Uveitis and Toxoplasmosis. Churchill, London, 1961.

PERRY, H., YANOFF, M., AND SCHEIE, H. G. Rubeosis in Fuchs heterochromic iridocyclitis. Arch. Ophthalmol. 93:337–339, 1975.

ZIMMERMAN, L. E. Ocular lesions of juvenile xanthogranuloma. Am. J. Ophthalmol. 60:1011–1035, 1965.

ZIMMERMAN, L. E., AND MAUMENEE, A. E. Ocular aspects of sarcoidosis. Am. Rev. Respir. Dis. 84:38–44, 1961.

3

IMMUNOLOGIC DISORDERS OF THE EYE

The immune system not only plays a crucial role in the body's defense against microorganisms, but humoral and cell-mediated immune responses may be injurious to the host (hypersensitivity, allergy). The superficial location of the eye constantly exposes it to numerous airborne antigens, such as pollens and bacterial proteins. The eyelids and conjunctiva are also often exposed to industrial chemicals and detergents, as well as drugs instilled into the eye in the therapy of ocular disease. All of these agents, although innocuous in themselves, often provoke allergic reactions in sensitized individuals.

The anatomic characteristics of the eye protect parts of it from the immunologic reactivity of the body. The normal cornea, lens, and anterior chamber lack lymphatics and blood vessels, and these features play a role in this regard. Because the anterior chamber is virtually isolated from immunologic reaction, many experimental biologists have made use of its superficial location to grow inoculated neoplasms. Heterografts are able to survive and proliferate in the anterior chamber and the growth of the neoplasm can be readily observed. Common substances causing allergic reactions of the eyelids include medications used for local ocular therapy, cosmetics, nail polishes, adhesive tapes, soaps, and hair sprays. The allergic reactions include atopic reactions, cell-mediated immune responses, and humoral immune responses. Repeated exposures to an antigen in a sensitized individual frequently accounts for the recurrent episodes.

ANAPHYLACTOGENIC REACTIONS

The anaphylactogenic reaction occurs immediately on exposure to the provoking agent. The response may involve a specific immunoglobulin (IgE) that mediates the release of vasoactive amines from mast cells in tissues and basophils in the blood. There is frequently an inherited predisposition to this reaction which is accompanied by an increased vascular dilation and permeability, and a cellular infiltrate of eosinophils. The reaction subsides quickly and is often accompanied by eosinophilia. The eyelids and conjunctiva are commonly affected. Hay fever is an example. In anaphylactogenic conjunctivitis (acute allergic conjunctivitis), chemosis and a serous exudate with numerous eosinophils are typical. Anaphylactogenic conjunctivitis may develop after the repeated local instillation of several different medications.

Vernal conjunctivitis (spring catarrh) is a chronic, recurrent, bilateral atopic conjunctivitis associated with an underlying tenacious, stringy, mucous discharge which contains an excess of eosinophils. The tarsal conjunctiva of the upper lid is thickened. Prominent gelatinous, flat topped papillae, which form a pattern like cobblestones, are characteristic (Fig. 2.4). Lymph follicle formation is minimal. The conjunctival goblet cells are increased in number. The subepithelial connective tissue, particularly in the tarsal conjunctiva, eventually becomes fibrosed. Recurrent episodes commonly occur with seasonal regularity in the warm months of the year, but not necessarily in the spring. Males are more frequently affected than females.

In allergic reactions of the eyelids, the palpebral skin becomes thickened and erythematous. Localized swelling of the superficial layers of the skin is usually sharply demarcated (urticaria).

CELL-MEDIATED IMMUNE RESPONSES

The immune response, known as delayed hypersensitivity (cellular hypersensitivity), occurs 12–48 hours after exposure to the antigen in an appropriately sensitized individual. It is mediated by sensitized circulating lymphocytes derived from the thymus (T cells). Cell-mediated immunity plays a role in diseases produced by numerous microorganisms, contact allergic dermatitis, allergic conjunctividites, and homograft (allograft) rejection, as well as autoallergies. Regardless of the nature of the antigen, the cellular infiltrate consists predominantly of lymphocytes, large mononuclear cells, and macrophages. In vascularized tissue, these cells are mainly around blood vessels. The severity of the lesion varies from a cellular infiltrate to severe fibrinoid necrosis of the vessel wall. Cell-mediated immune responses can be produced experimentally in the eye by the intraocular or intracorneal injection of antigen into appropriately sensitized animals. Delayed hypersensitivity can occur in the cornea. It is accompanied by a dense limbal infiltrate of plasma cells and their precursors.

An uncommon variety of keratoconjunctivitis that is probably a delayed hypersensitivity to several antigens, including some mycobacteria and staphylococci, is known as phlyctenular keratoconjunctivitis.

This entity is characterized clinically by a 1-3 mm wide nodule in the bulbar conjunctiva and contiguous cornea (phlyctenule) surrounded by a zone of hyperemia. The lesion consists mainly of densely packed small lymphocytes and may extend towards the central cornea followed by an ingrowth of blood vessels. The overlying epithelium may ulcerate.

Contact dermatitis may be caused by nickel or plastic in the frames of eyeglasses (spectacle dermatitis).

HUMORAL IMMUNE RESPONSES

Antibody production by lymphocytes (B cells) and plasma cells can occur in the ocular structures that normally contain lymphoid tissue, such as the conjunctiva. It may also occur in areas of the eye, such as the iris and ciliary body, that are infiltrated by immunologically competent cells.

Arthus Reactions

When antigens react with circulating complement-fixing antibodies (IgM or IgG) in the walls of blood vessels in the vascularized portions of the eye, thrombosis and fibrinoid necrosis of the blood vessels accompany a predominantly polymorphonuclear leukocytic infiltrate (local Arthus reaction). Numerous mononuclear cells and plasma cells are evident in the later stage in the evolution of the lesion. The phenomenon is well studied following the intraocular and subconjunctival injection of antigen into sensitized animals. It appears within about ½ hour and reaches its maximum intensity usually within several hours.

The Arthus reaction plays a role in experimentally produced immunologic uveitis and probably also in some forms of uveitis in man. The intravitreal inoculation of a foreign antigen into animals is followed soon by a partial loss of antigen from the eye and a concomitant appearance of it in the blood stream. Eventually, at about the 8th day, the circulating antigen disappears with the onset of immune elimination. Much of the antigen remains within the eye and slowly diffuses into the circulation. The phase of antigen elimination coincides with the appearance of an acute inflammatory reaction in the eye. Initially, hyperemia and edema of the uvea occur. Subsequently, there is an infiltrate of polymorphonuclear leukocytes followed by a progressive increase in mononuclear cells. Should the animal be sensitized by the systemic administration of the antigen before its injection into the vitreous, an immediate self-limited uveitis develops. This is followed by a subsequent reaction 7-10 days later. The initial reaction occurs in response to the antigen-antibody-complement complexes. After this, the intraocular antigen is thought to diffuse slowly out of the eye and to stimulate the immune system. The later response is believed to result from the migration of immunologically competent cells to the eye. It can be prevented by total body, but not ocular, x-irradiation. If the antigen is administered systemically many months after initial sensitization, a violent uveitis develops.

Because the cornea lacks blood vessels, the Arthus reaction does not occur there. However, circulating antibodies play a role in corneal hypersensitivity. After a soluble antigen is introduced into the cornea of a sensitized animal, it diffuses toward the corneoscleral limbus, while antibodies diffuse into the cornea. A ring of opacification develops within hours between the center of the cornea and the limbus where the antigens and antibodies react and precipitate ("immune ring," "Wessely ring") (Figs. 3.1–3.3). The zone of precipitation is analogous to that which occurs after antigens and antibodies diffuse toward each other in gels (Ouchterlony's diffusion system). The immune ring is composed of precipitated antigen-antibody-complement complexes and a pronounced infiltration of polymorphonuclear leukocytes that results from the local antigen-antibody reaction. All animals that develop the immune ring have detectable antibody in the serum. Should the animal not be sensitized, the immune ring may develop many days later, presumably as a result of sensitization to the intracorneally injected antigen. After

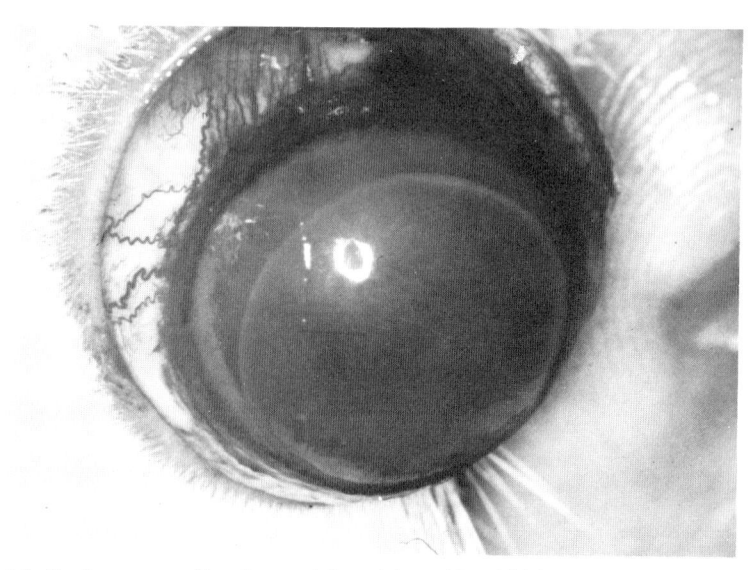

Figure 3.1. Bovine serum albumin was injected into this rabbit's cornea after the animal had been sensitized to the same protein. Note the opaque zone of precipitation ("immune ring") where the antigen and antibody met.

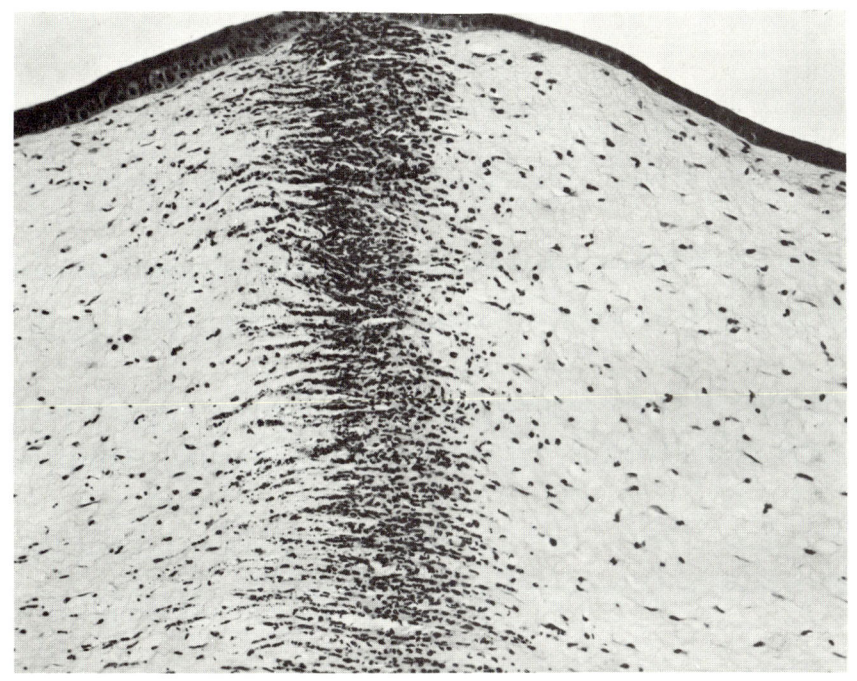

Figure 3.2. Numerous leukocytes, and especially neutrophils, accumulate at an "immune ring." (H & E, × 130)

the intracorneal injection of antigens into rabbits, antigen-antibody-producing cells are found in the homolateral preauricular lymph nodes, cornea, and uveal tract. Antibody production may be elicited by amounts of antigen which are inadequate to stimulate antibody production after systemic administration.

AUTOALLERGIES

A wide variety of autoantigens are restricted to particular tissues or cells. For instance, extracts of the lens contain, as determined by immunoelectrophoresis, at least 16 antigens. Such antigens have had little opportunity to impress their individuality upon the immune system, which is not normally exposed to them. Should the immune system become exposed to them, it reacts as if they were foreign antigens. Such a pathogenetic mechanism may play a role in at least two uncommon ocular conditions: sympathetic ophthalmia and phacoanaphylactic endophthalmitis.

Figure 3.3. The localization of fluorescein-labeled IgG at the site of the immune ring in this unstained tissue section is shown by fluorescence microscopy. ($\times$ 130)

Sympathetic Ophthalmia (Sympathetic Ophthalmitis)

A progressive, bilateral, diffuse granulomatous inflammation of the uveal tract occasionally follows a perforating accidental ocular injury with prolapse of uveal tissue, but rarely follows evisceration, intraocular melanoma, or intraocular surgery. The uveitis (sympathetic ophthalmia) develops in the originally injured eye ("exciting eye") after a latent period of more than 10 days (usually 4-8 weeks, sometimes many years). The uninjured eye ("sympathizing eye") becomes affected at the same time or shortly thereafter. In the absence of treatment, patients with sympathetic ophthalmia often become blind due to a protracted, relapsing inflammatory process within the eye that culminates in phthisis bulbi, but many patients have a mild course. Vitiligo and graying of the eyelashes sometimes occur. If enucleation of the "exciting eye" occurs within 7 days, the condition does not develop in the uninjured eye. Intraocular suppuration seems to prevent the condition.

In sympathetic ophthalmia, the uvea contains an infiltrate of lymphocytes, which may extend along the intrascleral canals to become extraocular. Histiocytes and giant cells are also present and characteristically contain phagocytosed melanin. Unlike some other granulomatous reactions, necrosis and suppuration are absent. Nodules containing reactive retinal pigment epithelium, epithelioid cells, and macrophages commonly appear between Bruch's membrane and the retinal pigment epithelium (Dalen-Fuchs' nodules).

It is widely believed that the individual becomes sensitized to uveal isoantigens that provoke an autoimmune reaction. Pigment phagocytosis has led to a long held belief that the melanin granule contains the offending antigen. Many patients with sympathetic ophthalmia exhibit positive delayed hypersensitivity reactions to intradermal uveal pigments in the early stages of the disease. After 2 weeks, the intradermal site of inoculation contains epithelioid and giant cells and phagocytosed pigment. Similar positive skin tests also occur in individuals without sympathetic ophthalmia. Inasmuch as sympathetic ophthalmia is extremely rare following nontraumatic conditions of the uvea in which pigment is dispersed, some investigators have doubted the autoimmune hypothesis and instead have postulated that the uveal tract may become invaded by yet unidentified microorganisms.

Phacoanaphylactic Endophthalmitis

A granulomatous inflammatory reaction in the anterior chamber of the eye centering around the lens or its remains, and even within the lens if the capsule is disrupted, is known as phacoanaphylactic endophthalmitis. It may occur in an eye with a traumatized or cataractous lens, or after the extraction of a cataractous lens. The reaction may occur spontaneously many months or years later in the contralateral eye if cataractous or injured. The cellular infiltrate varies considerably. It may consist mainly of neutrophils, eosinophils, lymphocytes, plasma cells, or macrophages. Epithelioid cells and multinucleated giant cells of the Langhans or foreign body type may be present. This rare reaction is thought to be due to sensitization by proteins that escaped from the lens. Antibodies to lens antigens can be provoked experimentally, and immunization with autologous lens material can produce a granulomatous reaction around the damaged lens. In the past, it was widely believed that the extruded lens material could also provoke an inflammatory reaction because of some toxic property of the lens (phacotoxic endophthalmitis, phacotoxic uveitis).

TISSUE TRANSPLANTATION

The transplantation of human corneal tissue is an established method of treating many corneal diseases (Fig. 3.4). These corneal homografts (allografts) frequently are successful despite immunologic differences between the donor and the recipient. Occasionally, the graft becomes

40   *The Eye*

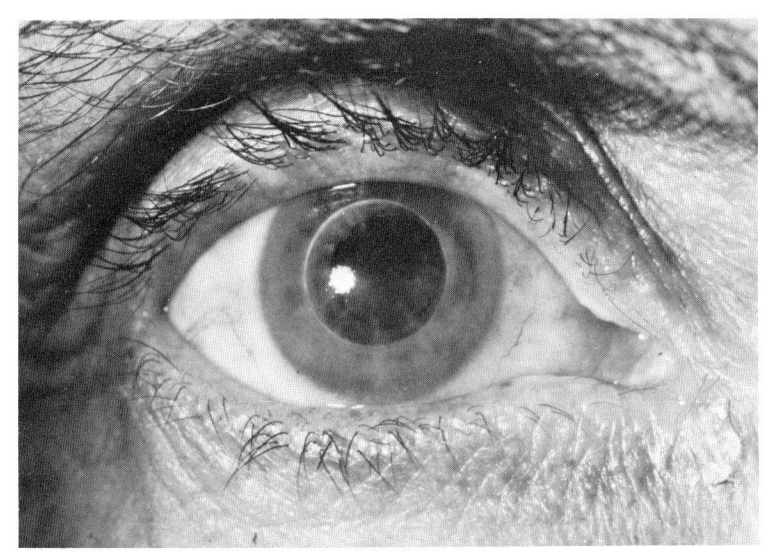

Figure 3.4. A corneal graft. A clear donor corneal graft has been surgically placed in a hazy recipient cornea. Note the faint white suture tracts about the graft-recipient junction.

opaque due to an immune rejection which usually occurs at least 2 weeks after the operation. Sensitized lymphocytes have been shown to play a role in graft rejection, although the possible contribution of humoral responses remains unclear.

Why the transplantation of corneal tissue between antigenically dissimilar individuals should ever be successful warrants consideration. Several factors play a role. Transplantation and tissue-specific antigens within a homograft are capable of providing an immunologic stimulus. Corneal homografts have relatively few antigens, especially transplantation antigens, as the tissue contains relatively few cells, particularly if the epithelium is denuded. Corneal homografts are populated frequently by cells of the host with time, but the cellular elements in the graft can persist indefinitely in experimental animals in the absence of rejection. The possibility of stromal mucopolysaccharides interfering with the access of transplantation antigens to immunologically competent cells has also been raised. Moreover, the lack of lymphatics and blood vessels in a normal cornea interpose a barrier between the antigens of the graft and the immunologically competent cells of the recipient. Assuming that the avascular recipient bed permits corneal homograft success, one would expect rejection in the presence of corneal vascularization. The accumulated clinical experience of ophthalmologists indicates that this does occur. Vascularization also plays a prominent role in homograft rejection of other tissues. Furthermore, vascularized corneas also contain lymphatics which drain into the regional lymph nodes. It is possible that antigens reach immunologically competent cells by this or other routes.

IMMUNOLOGICALLY RELATED DISEASES

Rheumatoid Arthritis

In rheumatoid arthritis, and occasionally in the absence of this disease, typical rheumatoid nodules, characterized by a central acellular eosinophilic area of fibrinoid necrosis of collagen surrounded by radial pallisades of elongated epithelioid cells, giant cells, plasma cells, lymphocytes, and eosinophils, may involve the sclera and episclera (brawny scleritis, annular scleritis) (Fig. 2.5). Diffuse rheumatoid scleritis produces marked thickening of the sclera. Discrete nodules in the anterior sclera may spread into the uvea and adjacent tissues and even ulcerate through the overlying conjunctiva. The necrotizing inflammatory scleral lesions may progressively perforate the sclera and result in uveal prolapse (scleromalacia perforans). These scleral lesions are usually anteceded by arthritis, but may precede joint involvement by as long as 10 years. The ocular lesions in rheumatoid arthritis are frequently bilateral although both eyes may not be affected simultaneously. In juvenile rheumatoid arthritis (Still's disease), bilateral uveitis and band keratopathy constitute a distinctive feature in about 10% of the cases.

Sjögren's Syndrome

Sjögren's syndrome consists of the triad of keratoconjunctivitis sicca, xerostomia, and enlargement of the parotid glands. It is associated with the destruction of the lacrimal and salivary glands and their infiltration by lymphocytes and plasma cells. The lymphoproliferation may become more generalized and involve other tissues. The disease usually affects middle-aged women. One-half the patients also suffer from rheumatoid arthritis or another connective tissue disease. Autoantibodies to lacrimal and salivary glands, hypergammaglobulinemia, rheumatoid factor, antinuclear antibodies, and other autoantibodies frequently occur.

Miscellaneous Conditions

The retinal and choroidal blood vessels may be involved in *periarteritis nodosa*. The necrotizing angiitis of this disease characteristically involves segments of arteries at which sites thromboses or aneurysms may develop. If the latter rupture, hemorrhage occurs. A circumscribed necrosis of the cornea and sclera, especially at the corneoscleral limbus, may occur, probably due to an occlusion of the anterior ciliary and scleral vessels. Other ocular lesions described in periarteritis nodosa include conjunctival edema, sensory retinal detachment, and hypertensive retinopathy.

Hypertensive retinopathy occurs in *systemic lupus erythematosus* and patients with this disease occasionally present with edema of the optic disc. Cotton-wool spots indicative of retinal ischemia often occur at the posterior pole of the fundus, even in the absence of vascular hypertension. Superficial and deep retinal hemorrhages occur secondary to venous and arterial occlusions. The characteristic facial rash of systemic lupus erythematosus may involve the eyelids. The ocular manifestations of *progressive systemic sclerosis* (generalized scleroderma) include involvement of the eyelids, cataracts, graying of the eyelashes (poliosis), and cotton-wool spots without vascular hypertension. A retinopathy with cotton-wool spots, as well as inflammation of the eyelid, sometimes occurs in *dermatomyositis*. In *hypersensitivity angiitis* (allergic vasculitis), a necrotizing inflammation of the smaller arterioles and venules of the eye and its adnexa may be involved together with vessels elsewhere in the body. The reaction is a sequel to hypersensitivity to a wide variety of antigens, including drugs.

Erythema Multiforme

Erythema multiforme is characterized by a severe bullous eruption of the skin and mucous membranes. The condition usually develops in children and young adults and can be precipitated by viruses and a variety of drugs including sulfonamides. Sharply demarcated erythematous patches of variable appearance appear asymmetrically in crops. They evolve into vesicles and bullae surrounded by varying degrees of erythema. Some lesions are hemorrhagic. The lesions may ulcerate, heal, scar, and give rise to pigmented or depigmented areas. Involvement of the conjunctiva often occurs (Stevens-Johnson syndrome). Scarring of conjunctival lesions commonly gives rise to adhesions between the tarsal and bulbar conjunctiva (symblepharon), trichiasis, and keratoconjunctivitis sicca. The eyelids may also be affected.

REFERENCES

ALLANSMITH, M. R., AND O'CONNOR, G. R. Immunoglobulins: structure, function and relation to the eye. Survey Ophthalmol. 14:367–402, 1970.

ANDERSON, L. G., AND TALAL, N. The spectrum of benign to malignant lymphoproliferation in Sjögren's syndrome. Clin. Exp. Immunol. 9:199–221, 1971.

ARONSON, S. B. The homoimmune uveitises in the guinea pig. Ann. N.Y. Acad. Sci. 124:365–376, 1965.

Aronson, S. B., Hogan, M. J., and Zweigart, P. Homoimmune uveitis in the guinea pig. Arch. Ophthalmol. 69:105–109, 203–207; 208–219, 1963.

Collins, R. C. Experimental studies on sympathetic ophthalmia. Am. J. Ophthalmol. 32:1687–1699, 1949.

Collins, R. C. Further experimental studies on sympathetic ophthalmia. Am. J. Ophthalmol. 36 (Part II):150–160, 1953.

Elliot, J. H. Immune factors in corneal graft rejection. Invest. Ophthalmol. 10:216–231, 1971.

Font, R. L., Yanoff, M., and Zimmerman, L. E. Benign lymphoepithelial lesion of the lacrimal gland and its relationship to Sjögren's syndrome. Am. J. Clin. Pathol. 48:365–376, 1967.

Foss, B. Experimental anaphylactic iridocyclitis. Acta Pathol. Microbiol. Scand. (Suppl.) 81:1–128, 1949.

Germuth, F. G., Maumenee, A. E., Senterfit, L. B., and Pollack, A. D. Immunohistologic studies on antigen-antibody reactions in the avascular cornea. I. Reactions in rabbits actively sensitized to foreign protein. J. Exp. Med. 115:919–928, 1962.

Henkind, P., and Gold, D. H. Ocular manifestations of rheumatoid disorders. Rheumatology 4:13–59, 1973.

Irvine, R. S., and Irvine, A. R., Jr. Lens-induced uveitis and glaucoma. Am. J. Ophthalmol. 35:177–186; 370–375; 489–499, 1952.

Khodadoust, A. A., and Silverstein, A. M. Studies on the nature of the privilege enjoyed by corneal allografts. Invest. Ophthalmol. 11:137–148, 1972.

Leskowitz, S., and Waksman, B. H. Studies in immunization. I. The effect of route of injection of bovine serum albumin in Freund adjuvant on production of circulating antibody and delayed hypersensitivity. J. Immunol. 84:58–72, 1960.

Movat, H. Z., Fernado, N. V. P., Uriuhara, T., and Weiser, W. J. Allergic inflammation. III. The fine structure of collagen fibrils at sites of antigen-antibody interaction in Arthus-type lesions. J. Exp. Med. 118:557–564, 1963.

Parks, J. J., Leibowitz, H. M. I., and Maumenee, A. E. A transient stage of suspected delayed sensitivity during the early induction phase of immediate corneal sensitivity. J. Exp. Med. 115:867–880, 1962.

Porter, R., and Knight, J. (Eds). Corneal Graft Failure. Associated Scientific Publishers, Amsterdam, 1973.

Rahi, A. H. S., and Garner, A. Immunopathology of the Eye. Blackwell Scientific Publications, Oxford, 1975.

Sevel, D. Necrogranulomatous scleritis. Am. J. Ophthalmol. 63:250–255, 1967.

Silverstein, A. M., and Zimmerman, L. E. Immunogenic endophthalmitis produced in the guinea pig by different pathogenetic mechanisms. Am. J. Ophthalmol. 48 (Part II):435–447, 1959.

Waksman, B. H., and Bullington, S. J. A quantitative study of the passive Arthus reaction in the rabbit eye. J. Immunol. 76:441–453, 1956.

Waksman, B. H., and Bullington, S. J. Studies of arthritis and other lesions induced in rats by injection of mycobacterial adjuvant III. Lesions of the eye. Arch. Ophthalmol. 64:751–762, 1960.

4

VASCULAR DISORDERS OF THE EYE

The retina is in a part of the body where the small blood vessels can be observed directly, and where their circulation can be studied in the living individual. The latter is aided by fluorescein angiography in which a fluorescent dye is injected intravenously. Using an appropriate set of filters, it is possible to photograph only the fluorescent dye in the retinal blood vessels, and to screen out all of the other intraocular structures. In fluorescein angiograms, it is possible to visualize all retinal vessels including normal retinal capillaries (approximately 5 μ in diameter) (Fig. 4.1). Vascular pulsations occur in the retina synchronously with cardiac contractions. Normally, this expansile pulsation is observed more readily in the retinal veins than in arteries. Abnormal vascular pulsations occur in a variety of conditions. For example, a cardiac arrhythmia may reflect itself in an abnormal pulsatile rhythm. Enhanced arterial pulsation occurs if there is high pulse pressure, as in aortic insufficiency, or if the intraocular pressure is elevated in relation to the diastolic blood pressure. If the central retinal venous pressure is elevated as with an impending central retinal vein thrombosis or papilledema, the pulsation is no longer evident. A variety of other vascular abnormalities can be observed on funduscopic examination in living patients. These include variations in the caliber of vessels and alterations in the appearance or tortuosity of the vessels. Cyanosis may be apparent on funduscopy, especially in patients with cyanotic congenital heart disease.

BLOOD VESSELS

Several branches of the ophthalmic artery supply the eye, optic nerve, and orbital structures (Fig. 4.2). The central retinal artery penetrates the optic nerve a short distance from the globe and runs in its center to enter the globe at the optic disc. Branches of the central retinal vein converge at the optic disc. Both the central retinal artery and vein traverse the subarachnoid space as they pass between the optic nerve and the orbital tissue.

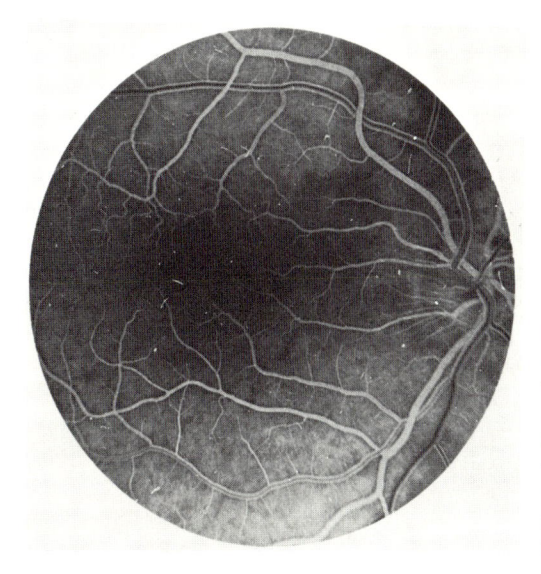

Figure 4.1. Normal fluorescein angiogram. Fluorescein dye fills the retinal arterioles giving them a whitish appearance. Return of flow is beginning in the retinal venous system showing a laminar pattern in the larger veins. Fluorescein dye does not leak out of normal retinal vessels.

The larger vessels in the retina are limited to the nerve fiber and ganglion cell layers. The inner portion of the retina is supplied by branches of the central retinal artery, while the outer retina receives nutrients by way of a fenestrated capillary network in the inner portion of the choroid—the choriocapillaris. Because of this dual blood supply to the retina, an impaired circulation in the choriocapillaris interferes with the nutrition of the outer retina, while an impaired retinal circulation affects the inner retina. The choroid, the most vascularized part of the eye, is supplied by branches of the ophthalmic artery that penetrate the sclera. Vortex veins drain the globe and enter the superior and inferior ophthalmic veins which communicate with the cavernous sinus and veins of the face—an anatomic fact that is important with regard to cavernous sinus thrombosis and carotid cavernous fistulae. The cornea, vitreous, and certain parts of the retina lack blood vessels.

Types of Capillaries

Several types of capillaries occur in the eye. In the retina, the adjacent endothelial cells of the capillaries are bound together by a band which consists of a zonular occludens adjacent to a zonular adherens that encircles the circumference of the cell. The endothelial cells are surrounded by a thick basement membrane which also surrounds the retinal pericytes (mural cells, intramural pericytes). The endothelial cell-pericyte complex is surrounded by glial cells and does not become

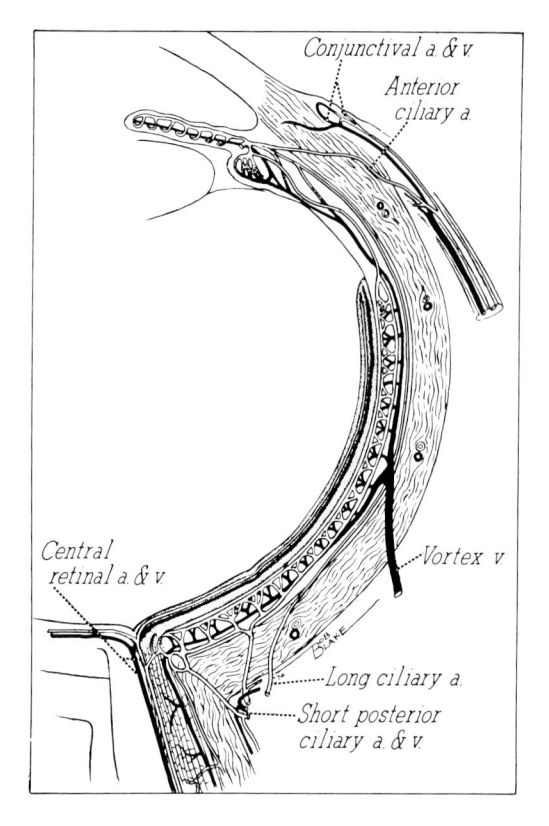

Figure 4.2. The blood supply and venous drainage is shown schematically.

more permeable in response to vasoactive amines such as histamine. These anatomic features prevent large molecules from passing extravascularly into the retina (the *blood retinal barrier*). Normally, the retinal capillaries possess an equal number of pericytes and endothelial cells. Vasoactive substances readily separate junctions between the endothelial cells of capillaries in the extraocular muscles, iris, and other ocular tissues. The choriocapillaris typifies yet another type of capillary. It possesses fenestrations and gaps between the vascular lumen and the extracellular space. Such a vascular system aids in the transport of nutrients toward the outer retina. The ciliary body also has fenestrated capillaries. Arteries in the iris have an unusually thick acellular wall with numerous collagen lamellae and little muscle. They pass regularly in a corkscrew manner toward the pupil.

HEMORRHAGE

The extravasation of blood may follow the traumatic severance of blood vessels, be part of a generalized bleeding diathesis, follow venous obstruction, or be associated with specific abnormalities of the eye. Important structures may be destroyed. A small hemorrhage in the macula may cause severe visual impairment. There is insignificant blood loss with ocular hemorrhages, and hemorrhagic shock does not ensue. Once hemorrhage has occurred, the debris is often phagocytosed and the hemoglobin is converted to hemosiderin. It may remain in macrophages for years, as after central retinal vein thrombosis. A rusty brown color may be evident, especially if there has been repeated bleeding.

Hemorrhages may be localized to specific areas of the eye. Their location often depends upon their cause and is hence of diagnostic significance. Conjunctival petechiae sometimes follow anoxia, or may follow severe bouts of coughing, although they often arise spontaneously. Conjunctival hemorrhages, as a rule, remain in the subepithelial tissue of the conjunctiva and do not extend into the cornea because of the unusual close apposition of the corneal epithelium to its underlying substantia propria. Retinal hemorrhages occur in a number of conditions, including hypertension, central retinal vein thrombosis, and diabetes mellitus. The shape of hemorrhages within the retina is dependent upon their location. Blood in the nerve fiber layer spreads between the axons and presents a flame-shaped appearance on funduscopy (Fig. 4.3). Deep retinal hemorrhages, on the other hand, are often round. Hemorrhage between the retinal pigment epithelium and Bruch's membrane may appear as a well circumscribed dark mass and clinically simulate a neoplasm.

Choroidal hemorrhages may be massive, especially in elderly hypertensive patients after accidental or surgical perforation of the globe.

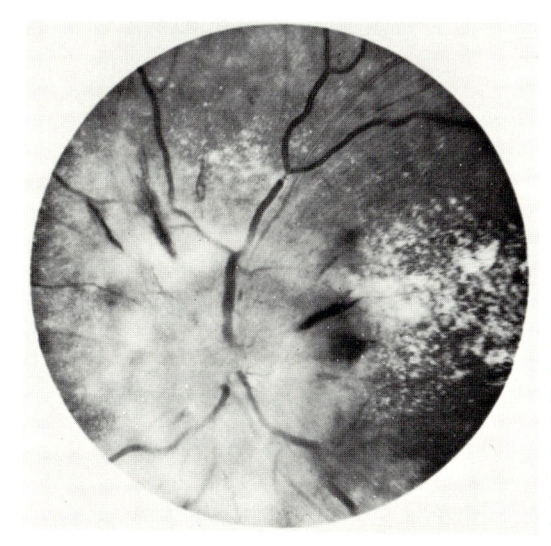

Figure 4.3. Flame-shaped retinal hemorrhages. Blood in the nerve fiber layer in a case of optic neuritis spreads between the axons and presents a linear appearance. The nerve head is swollen, its margins are indistinct, and there are exudates in the surrounding retina.

They detach the choroid and displace the retina, vitreous, and lens and sometimes expel them through the wound ("expulsive hemorrhages"). When present in the vitreous or the anterior chamber, the blood may gravitate to the most dependent part (Fig. 4.4). Blood in the anterior chamber (hyphema) is a common complication of ocular contusions. Hyphema also may accompany and sometimes be the presenting clinical feature of several ocular diseases, including retinoblastoma and melanoma. It also frequently results from bleeding of friable new vessels that form on the anterior surface of the iris (rubeosis iridis) in certain conditions. If there is relatively little blood in the anterior chamber, it may drain by way of the trabecular meshwork. Particularly in the presence of massive or recurrent hyphema, clots may form in the anterior chamber and become organized into granulation tissue. When this occurs, adhesions develop between the iris and the adjacent structures, and glaucoma may ensue secondary to the obstruction of the aqueous outflow channels. The cornea may become markedly discolored (corneal blood staining) by blood products when hyphema occurs in the presence of glaucoma. Hemorrhage in the vitreous commonly originates in the retina or choroid, especially after laceration of blood vessels. Subhyaloid and vitreous hemorrhages sometimes complicate a subarachnoid hemorrhage (Terson's syndrome). Blood in the vitreous is generally absorbed slowly. If the hemorrhage recurs or absorption is delayed, it becomes organized with scar formation and vascular invasion.

MICROANEURYSMS

Capillary microaneurysms seem to be restricted to the retina. These saccular pouchings of the capillary are 15–100 μ in diameter. They appear on funduscopic examination as minute dots and are seen more easily on fluorescein angiograms (Fig. 4.5). With this technique, they are evident around areas of nonperfused capillary closure and seem to arise from both the arterial and venous ends of capillaries, but especially from the venous side. Microaneurysms are not on closed vessels but on those with blood flow. In addition to saccular aneurysms, generalized or fusiform dilations of other capillary loops occur. There are kinks in some aneurysms. Microaneurysms are one of the commonest manifestations of retinal vascular disease. They occur in diabetes mellitus, central retinal vein occlusion, sickle cell disease, leukemia, the aortic arch syndrome, malignant hypertension, macroglobulinemia, multiple myeloma, severe anemia, other conditions, and rarely in otherwise clinically normal retinas.

Vascular Disorders 49

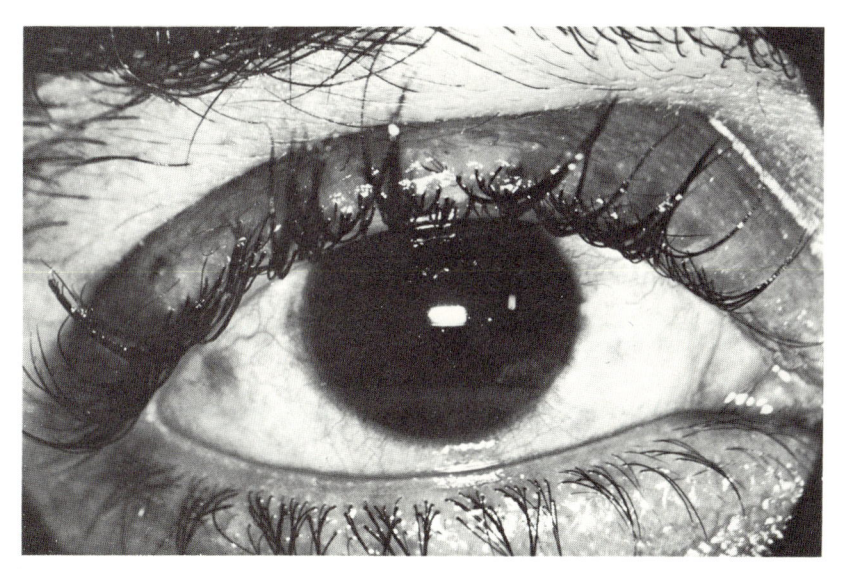

Figure 4.4. Hyphema. Red cells have layered out in the inferior aspect of the anterior chamber after blunt trauma to the eye.

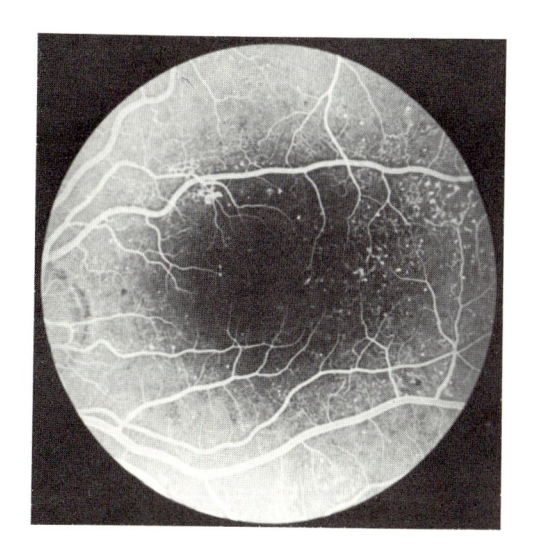

Figure 4.5. Fluorescein angiogram showing multiple microaneurysms in a case of diabetic retinopathy.

The pathogenesis of microaneurysms is still unsettled. A common factor in all conditions associated with them is retinal small vessel occlusive disease. The smaller and presumably earlier aneurysms have thin walls and are lined by an endothelium (Fig. 4.6).

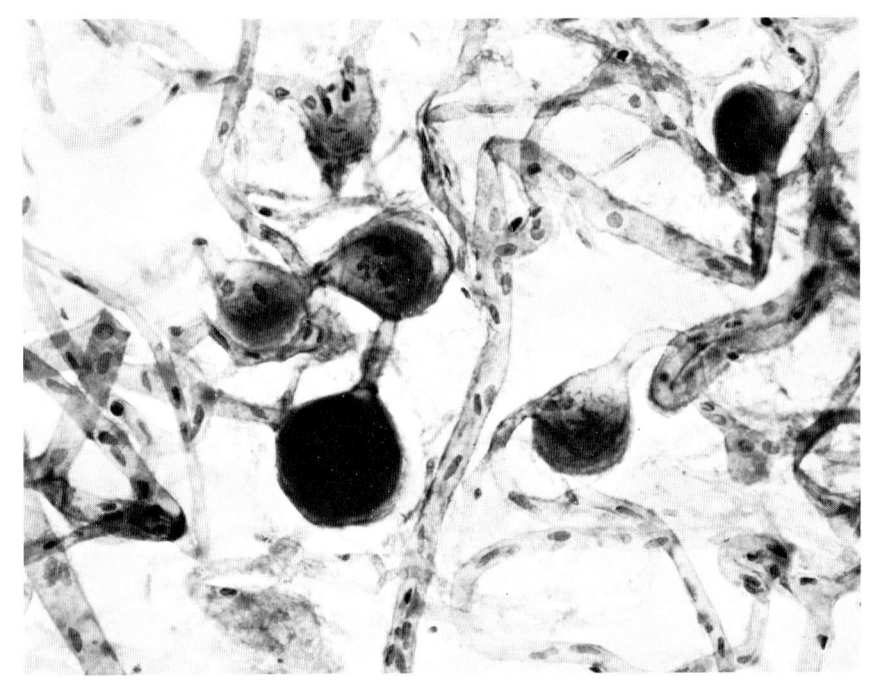

Figure 4.6. Numerous microaneurysms are present in this flat preparation of the retina. (periodic acid-Schiff, × 250)

NEOVASCULARIZATION

Neovascularization of the Retina

When areas of the retina have been deprived of their circulation, newly formed vessels sometimes form in the reparative phase (Fig. 4.7). New capillary formation occurs initially within the retina and arises from the venous side of the circulation with new vessels arising along the inner surface of the retina. Eventually, the vessels penetrate through the inner limiting membrane of the retina and extend in the plane between this membrane and the posterior surface of the vitreous. Eventually, they invade the vitreous. An effect of this is detachment of the retina by virtue of the fibroblastic tissue accompanying the invading newly formed blood vessels (retinitis proliferans). The newly formed vessels lack pericytes, are fragile, readily bleed, and are permeable to large molecules as evidenced by fluorescein angiography. The progression of new vessel formation may become spontaneously arrested. Ashton has suggested that retinal ischemia provokes the formation of a metabolite which stimulates vascular endothelial proliferation and that this factor diffuses into other parts of the eye including the vitreous, aqueous, and iris. The basic requirements for neovascularization of the retina seem to be the presence of living cells to ensure active metabolism, low oxygen tension to promote anaerobic metabolism, and a poor venous drainage to permit the accumulation of the vasoformative factor.

Rubeosis Iridis

A neovascular response of marked clinical significance is known as rubeosis iridis. It is characterized by the growth of a layer of fibroblasts and blood vessels along the anterior surface of the iris (Fig. 2.3). Rubeosis iridis occurs in diabetes mellitus, central retinal vein occlusion, retinoblastoma and other ocular neoplasms, carotid ischemia, long standing retinal detachment, carotid cavernous fistula, and other conditions. The feature common to most conditions with rubeosis iridis is reduced retinal blood flow. New vessels first appear on the anterior surface of the iris, usually at the pupillary margin, but they can arise at any point. Rubeosis iridis is often the death knell of any eye in which it appears. The meshwork of new vessels may cover the pupillary zone of the iris and often occludes the angles, resulting in a secondary closed angle glaucoma. These vessels are extremely prone to hemorrhage and hyphema frequently ensues. The fibrovascular membrane may result in ectropion uveae, peripheral anterior synechiae, and sometimes posterior synechiae. Rubeosis iridis has been produced experimentally in the rabbit by the obliteration of both of the long posterior ciliary arteries.

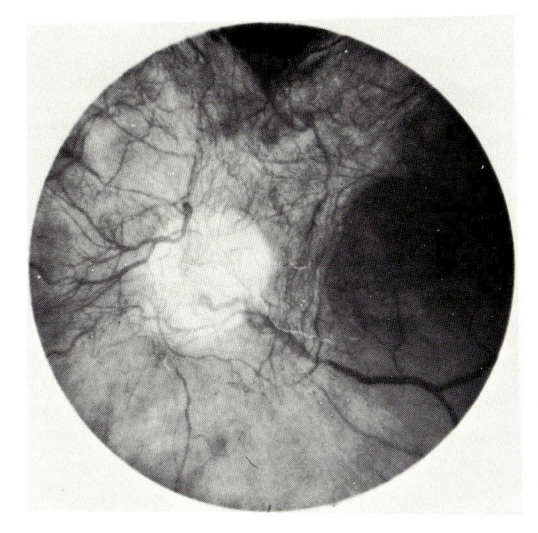

Figure 4.7. Retinal neovascularization. An irregular network of new blood vessels extends outward from the region of the optic nerve onto the surface of the retina and into the vitreous cavity.

Corneal Vascularization

An intriguing question is why the normal cornea is avascular and yet becomes vascularized in certain pathologic states (Fig. 4.8). Although this question remains unsettled, there is evidence that corneal vascularization depends upon the presence of one or more diffusible factors capable of stimulating directional capillary growth which the normal avascular cornea lacks. Corneal vascularization is usually preceded by an inflammatory cell infiltrate.

EXUDATES, TRANSUDATES, AND INSUDATES

An excessive accumulation of fluid in the ocular tissue spaces may occur due to a variety of causes. As in other parts of the body, it may result from an increased vascular hydrostatic pressure, a decrease in plasma proteins, or an increased capillary permeability. Lymphatic obstruction plays a negligible role in edema of ocular tissues as it is only the eyelids and conjunctiva that have a lymphatic drainage.

Different terms have been applied to an excessive accumulation of fluid in the tissue spaces depending upon the location and the composition of the fluid. The fluid may contain little protein and a few cells (edema, transudates), or much protein and/or cells (exudate). Plasma constituents, like protein and lipid, may infiltrate into the vacular wall (insudate). A transudate results from an increased production or impaired absorption of the extracellular fluid. Exudates develop when the integrity of the vessel wall becomes impaired and permits the extravascular permeation of constituents that normally reside in the vascular compartment. With time, the fluid within the exudate becomes reabsorbed leaving a protein-rich and often lipid-rich residue.

Transudates

Edema of the Optic Disc (Papilledema)

Edema of the optic disc may be unilateral, but is most often bilateral (Fig. 4.9). It may result from increased intracranial pressure due to any cause, malignant hypertension, obstruction to the venous drainage of the retina and optic nerve as with expanding lesions of the orbit, cavernous sinus thrombosis, and carotid cavernous fistula, intrathoracic venous obstruction, chronic respiratory disease, acute glaucoma, acute inflammation of the optic nerve, decreased intraocular pressure (hypotony), and miscellaneous conditions in which the pathogenesis of the edema of the optic disc is poorly understood (severe anemia, polycythemia vera).

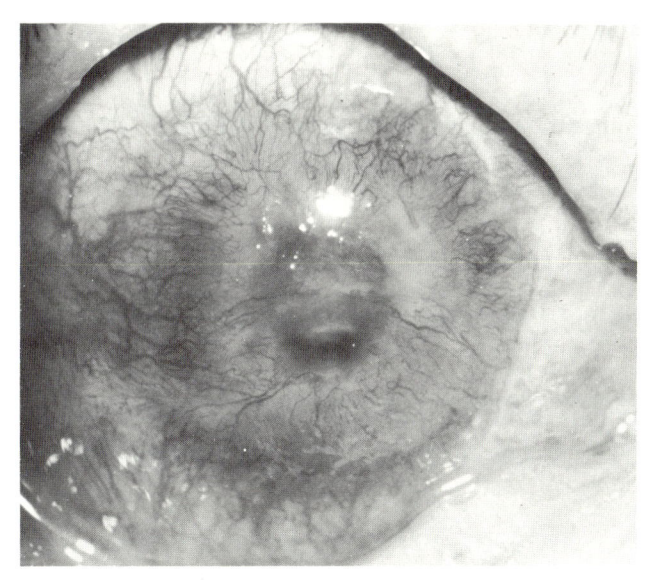

Figure 4.8. Corneal vascularization. Blood vessels have invaded the normally avascular cornea.

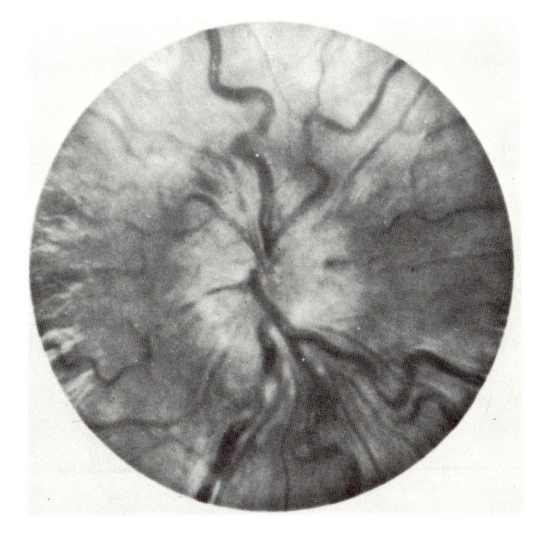

Figure 4.9. Papilledema with blurring of the disc margins, as well as congestion and tortuosity of the retinal vessels. Exudates and hemorrhages are also present within the retina.

When viewed with the ophthalmoscope, the appearance of the optic disc varies with the degree of edema. The margins of the disc become blurred, first at the superior and inferior disc margins, then the nasal side, and eventually the temporal side. There is usually an increased redness of the disc. The physiologic cup disappears, the optic nerve becomes elevated, and gray streaks occur along the retinal vessels. Edema may spread toward the macula. Venous and capillary distention occur. There is an anterior displacement of the retinal vessels. The retinal veins become dilated and lose their pulsations. There may be hemorrhaging on the disc surface and in the retinal nerve fiber layer. It may extend into the vitreous. With marked increased intracranial pressure, the funduscopic appearance of the optic disc is sometimes identical to that of a central retinal vein occlusion. An enlargement of the blind spot may be detected clinically.

Optic neuritis may be indistinguishable funduscopically from papilledema due to other causes, but is usually unilateral and often accompanied by pain (especially on movement of the eye) and sudden loss of vision in that eye.

In tissue sections, the swollen optic nerve head protrudes forward into the vitreous. A subretinal serous exudate commonly buckles the outer retina and separates the adjacent peripapillary sensory retina from the retinal epithelium (Fig. 4.10). Demyelination and axonal degeneration may succeed the optic atrophy, and eventually gliosis of the nerve occurs.

Several structural characteristics of the optic nerve are believed to play a crucial role in the pathogenesis of papilledema due to increased intracranial pressure. The subarachnoid space surrounding the optic nerve is an extension of that around the brain with which it is generally in direct communication. Hence, the intracranial pressure is usually transmitted around the optic nerve. A direct communication between the intracranial subarachnoid space and that around the optic nerve is essential for papilledema to develop. The central retinal vein draining the retina runs in the axial portion within the optic nerve for about 8–15 mm before leaving the nerve and crossing the meninges. Should the intracranial pressure become sufficiently elevated, it may occlude the central retinal vein at a site where it traverses the subarachnoid space. However, venous obstruction alone clearly does not cause papilledema, as it is not a feature of congestive cardiac failure and superior vena cava obstruction. The optic disc possesses a rich arterial blood supply derived from branches of the central retinal artery, choroidal arteries, and an arterial circle within the rigid sclera that surrounds the nerve head (arterial circle of Zinn-Haller). The relative differences between the arterial and venous pressure seem to be relevant to the development of papilledema.

The degree of edema varies and is often proportional to the increase of intracranial pressure. All patients with marked increased intracranial pressure do not develop papilledema. In some instances, this may result

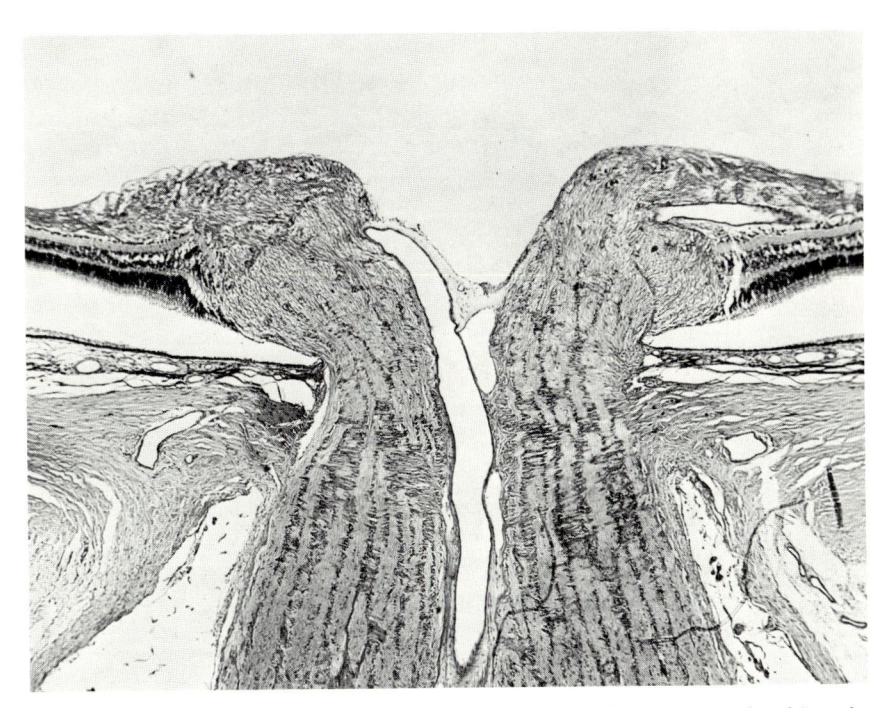

Figure 4.10. Edema of the optic disc. Note the protrusion of the optic nerve head into the vitreous. (H & E, × 25)

from the subarachnoid space surrounding the optic nerve not communicating with the intracranial portion because of a tight encirclement by its surrounding meninges, particularly within the optic foramen. Also, certain intracranial masses, such as meningiomas, may compress the optic nerve directly and obstruct the optic foramen resulting in increased intracranial pressure with unilateral papilledema with the compressed side often developing optic atrophy (Foster Kennedy syndrome). The possibility of lymphatic obstruction causing papilledema is readily excluded by the absence of lymphatics in the optic nerve.

Studies related to the pathogenesis of papilledema have been performed by several investigators, and notably by Hayreh. If the optic nerve is first ligated proximal to the exit of the central retinal vein, papilledema does not develop with increased intracranial pressure, but, if the nerve is ligated closer to the globe, papilledema does develop. Papilledema can be prevented if the elevated intracranial pressure is prevented from reaching the globe by sectioning the dural sheath of the optic nerve. Low intraocular pressure facilitates the experimental production of papilledema with increased intracranial pressure. In malignant hypertension, papilledema also seems to be related to increased intracranial pressure, although it does not necessarily parallel the cerebrospinal pressure. Papilledema due to hypotony presumably results from an impaired absorption and drainage of interstitial fluid from the eye. Polycythemia vera can cause papilledema because of central retinal vein occlusion. Spinal cord tumors are sometimes associated with papilledema, possibly due to the elevated protein in the cerebrospinal fluid obstructing its drainage. With chronic respiratory disease, papilledema may result from venous obstruction at the level of the lungs, generalized hypoxia, and/or an associated central retinal vein thrombosis, perhaps as a sequel to secondary polycythemia.

Edema of the Conjunctiva and Eyelids

Conjunctival edema (chemosis) accompanies conjunctivitis and also can occur with stasis within orbital or ocular veins or lymphatics. The loosely arranged subcutaneous tissue of the eyelids accounts for the pronounced swelling that frequently accompanies edema of the eyelids (Fig. 4.11).

Retinal Transudates

Retinal edema notably follows central retinal vein occlusion and certain acute inflammatory lesions of the retina. Ophthalmoscopically, it appears as a pale white area with the retinal vessels barely visible. Microscopically cystoid spaces occur in the outer plexiform layer of the retina. The fovea, which lacks the inner layers of the retina, is not as extensively involved as other parts of the retina.

Exudates

Subretinal Exudates

As a rule, when the sensory retina separates from the retinal pigment epithelium, an exudate with a high protein content accumulates between the separated portions of the retina. A subretinal exudate also may follow the increased vascular permeability that accompanies melanomas, hemangiomas, and other diseases.

In young children, a massive subretinal exudate (Coats' lesion) may develop with congenital vascular abnormalities of the retina, birth trauma, the battered baby syndrome, or nematode endophthalmitis. Exudates within the retina are commonly associated with the subretinal exudate. Usually, one eye is affected and Coats' lesion needs to be differentiated clinically from a retinoblastoma. Cholesterol crystals and

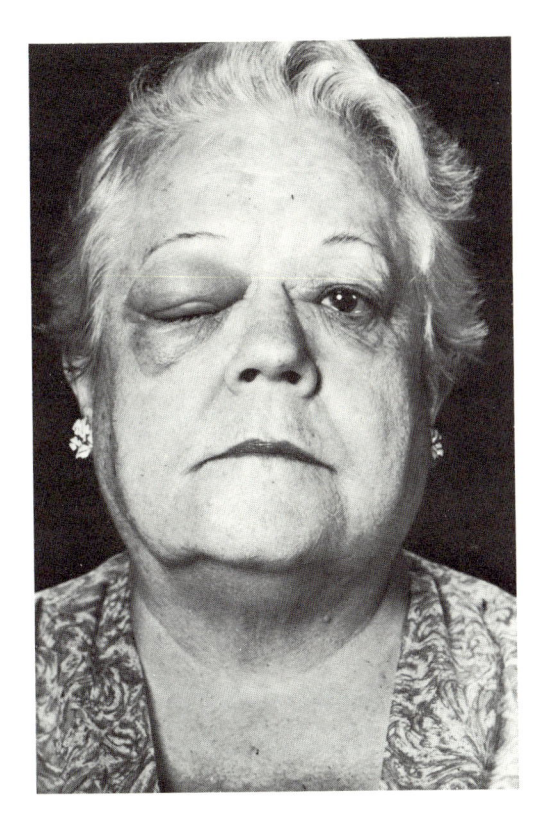

Figure 4.11. Edema of the eyelids in a resolving case of orbital cellulitis.

lipid-filled macrophages are prominent constituents of the subretinal exudate. It is thought that the exudate follows an increased vascular permeability to plasma and that it becomes partially inspissated by the absorption of fluid resulting in the precipitation of cholesterol and neutral fat out of solution. The cholesterol crystals frequently become surrounded by foreign body giant cells.

Retinal Exudates

True retinal exudates have been variably referred to as fatty exudates, hard exudates, waxy exudates, and chronic edema residues (Figs. 4.3, 4.12). They are situated mainly in the outer plexiform layer, hence the term deep exudates. Around the macula, the exudates often lie between the radial fibers of Henle's layer (axons of the photoreceptor cells at the macula) and show a stellate configuration (macular star). Retinal exudates are common in diabetes mellitus, Coats' lesion, central retinal vein thrombosis, and retinal angiomas. Depending upon the age of the exudate, the extravasated material may lie extracellularly within the

58 The Eye

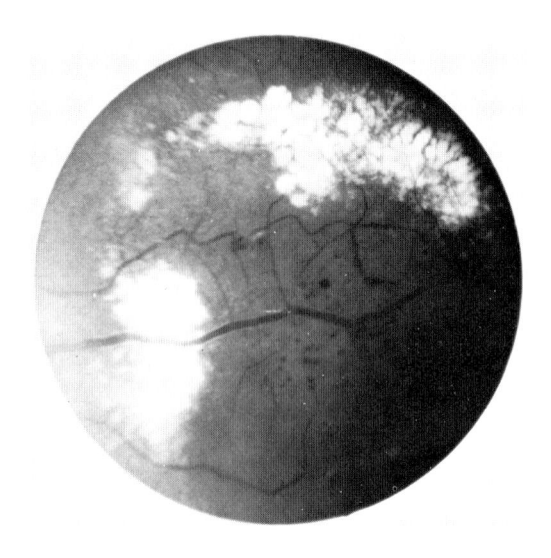

Figure 4.12. Retinal exudates arranged in a semicircular pattern around an area of abnormal retinal blood vessels and microaneurysms.

retina or partially within macrophages (Fig. 4.13). Like exudates elsewhere, they vary in composition. By light microscopy, the exudates are eosinophilic, periodic acid-Schiff-positive, and often contain neutral fat. The lipid-rich residue corresponds to what ophthalmologists call the hard waxy exudate.

Insudate

A considerable body of evidence provided by histochemical, immunofluorescence, and electron microscopic methods lends support to the view that hyaline arteriosclerosis and the vascular lesions in diabetes result from the infiltration of plasma constituents like protein, lipid, and mucopolysaccharides into the vascular wall. The mechanisms responsible for this infiltration of plasma components are not clear, but mechanical factors, such as an increased hydrostatic pressure, and increased vascular permeability, have been proposed. Hypertension plays a role in the genesis of hyaline insudation of arterioles. Varying degrees of arteriolar hyalinization also occur in individuals without hypertension. Blood proteins also enter the arteriosclerotic plaques of large arteries.

HYPEREMIA (VASODILATION)

Hyperemia follows an increased inflow, as in the acute inflammatory response, and a decreased outflow, as with venous obstruction. Vasodilation that begins in one vessel may readily spread to others because of anastomotic channels between blood vessels. This explains the commonly encountered concomitant conjunctival and ciliary vascular congestion.

A red eye may result from acute inflammation of the eye, as with conjunctivitis and acute iridocyclitis. The superficial conjunctival arterial plexus commonly dilates in keratitis, while the deep pericorneal conjunctival arteries become hyperemic in iritis and angle closure glaucoma. In conjunctivitis, the superficial conjunctival vessels are constricted by the topical application of epinephrine. Because of their superficial location, they are readily movable over the underlying sclera. The posterior conjunctival blood vessels are dilated, especially in the fornices, in inflammation of the bulbar conjunctiva. The blood vessels diminish in size toward the corneoscleral limbus. On the other hand, in acute iridocyclitis there is a pericorneal halo and dilated deep ciliary vessels. The deep location of these vessels accounts for their lack of mobility with the overlying conjunctiva, their dull red to purple color, and their failure to constrict after topical instillation of epinephrine. The red eye of acute glaucoma is hard to palpation and has a steamy cornea as well as a semidilated fixed pupil.

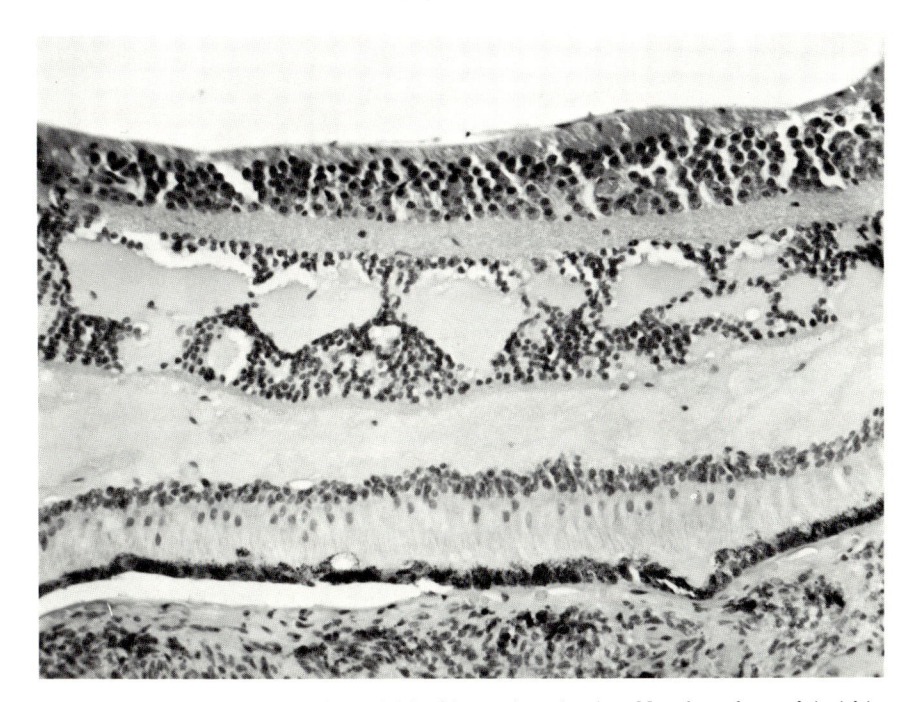

Figure 4.13. Exudates are evident within this portion of retina. Note how the nuclei within the inner nuclear layer of the retina have become displaced. (H & E, × 160).

OCCLUSOVASCULAR DISEASE

Vascular occlusion may result from thrombosis, embolism, stenosis (as occurs in atherosclerosis), vascular compression, intravascular sludging or coagulation (e.g. sickle cell disease, disseminated intravascular coagulation), or vasoconstriction (e.g. hypertensive retinopathy and migraine). Thrombosis may accompany diseased blood vessels as in syphilis, giant cell arteritis, and thromboangiitis obliterans.

Embolic Disease

Emboli arising from the heart or major vessels such as the carotid arteries may lodge in the eye. Diseases which predispose to ocular emboli are bacterial endocarditis, cardiac mural thrombi, calcified cardiac valves, atrial myxomas, and ulcerated atheromatous plaques. Small emboli, such as cholesterol crystals (Figs. 4.14, 4.15), commonly obstruct branches of the retinal arteries and may be evident on funduscopic examination at points of bifurcation of retinal arteries. The most frequent site of embolic obstruction of the central retinal artery is in the area of the lamina cribrosa. Here the central retinal artery and vein are fixed to each other, as well as to surrounding connective tissue by dense collagenous and elastic tissue. The lumen of the artery is narrower here than in its orbital portion, an anatomic feature that is thought to account

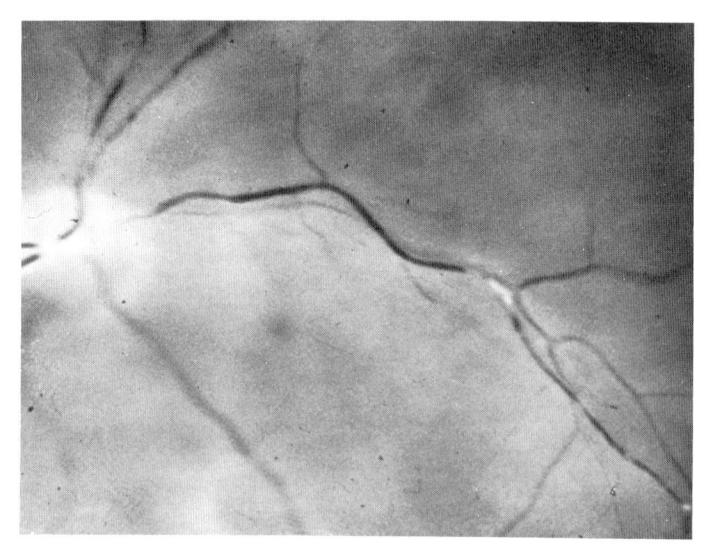

Figure 4.14. A bright plaque of embolic cholesterol crystals is evident at a bifurcation of a retinal artery.

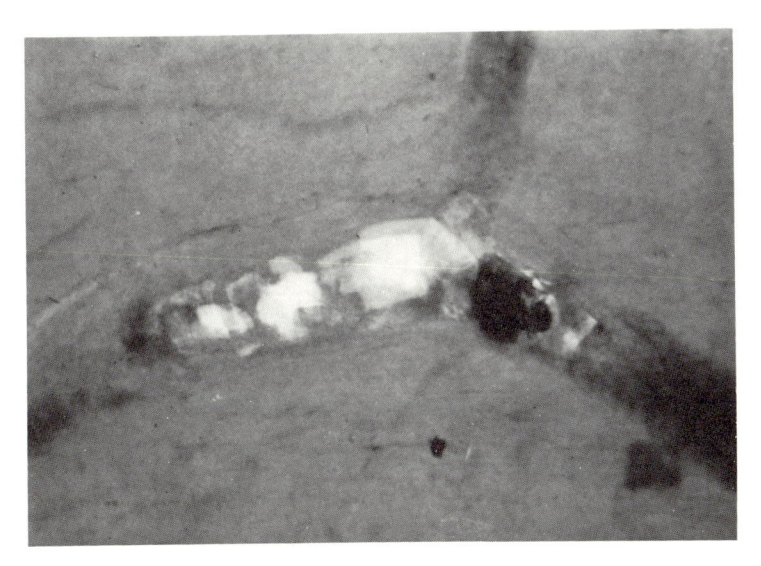

Figure 4.15. This photograph of the same retina as in Fig. 4.14 shows numerous crystals of cholesterol under polarized light. ($\times$ 180)

for the greater frequency of occlusion at or near this location. Fat emboli and other small emboli may reach the eye by way of collateral blood vessels without passing through the lungs. Fat emboli occur especially after fractures of long bones and their detection within the eye may aid in the clinical diagnosis of fat embolization. Air emboli may occur after sudden barometric decompression, after surgical procedures, or after accidental injuries to the neck or thorax.

The effect depends upon the size of the vessel involved and the degree of vascular occlusion. When only a small vessel is occluded, the collateral circulation may be adequate enough to compensate. Small emboli may not interfere with retinal function as they often do not completely occlude the retinal vessels. Infected emboli may cause a focus of ocular infection. Fibrin and platelet emboli may cause transient occlusive symptoms and leave no clinical trace of their presence.

Vasoconstriction

Vasoconstriction of the retinal vessels may be observed clinically in hypertensive retinopathy and migraine. A generalized arteriolar vasoconstriction accompanies a marked increase in systemic blood pressure. Because the retinal vessels lack an innervation, these changes in the caliber of the retinal vessels result from locally controlled vasomotion.

Effects of Ischemia

Extremely white fluffy patches that resemble cotton (cotton-wool patches, "soft exudates") are frequently observed on ophthalmoscopic examination in patients with retinal ischemia (Fig. 4.16). These lesions are generally round in shape and seldom wider than the optic disc. Cotton-wool spots are observed in diabetic retinopathy, retinal occlusive vascular disease, malignant hypertension, pernicious anemia, collagen disease, and after central retinal vein occlusion. They are not exudates, but consist of aggregations of varicose swollen axons in the nerve fiber layer of the retina (Fig. 4.17). The individual swollen axons appear in tissue sections as spherical or pyriform structures (about 25 μ in diameter) with degenerated cytoplasmic organelles. Because of a superficial resemblance to cross-sections of cells, the abnormal axons have been referred to as cytoid bodies. The cotton-wool spot may possess a red halo due to surrounding hemorrhage.

Ashton and his colleagues have investigated the ultrastructure and pathogenesis of the experimentally produced cotton-wool spots which develop as a direct response to arteriolar obstruction. The lesions are potentially reversible depending on the duration of the circulatory failure and the capacity of the anastomotic channels to develop. The degenerated cytoplasmic organelles within them consist of numerous mitochondria and other dense bodies related to the lysosomal system.

Retinal ischemia deprives tissue of oxygen and other essential metabolites. The neurons of the retina, like those in the rest of the nervous

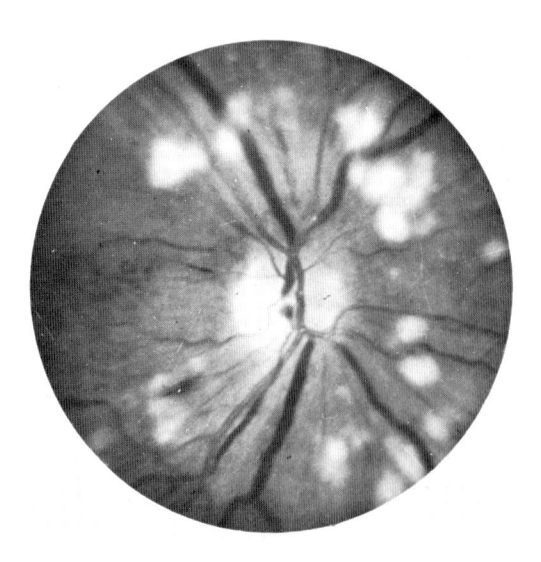

Figure 4.16. Cotton-wool spots. These white patches, in the superficial retina located here around the optic nerve, represent areas of swollen axons in the nerve fiber layer. They are so-called because of their resemblance to what is termed "cotton" in the United States and "cotton-wool" in most English-speaking countries.

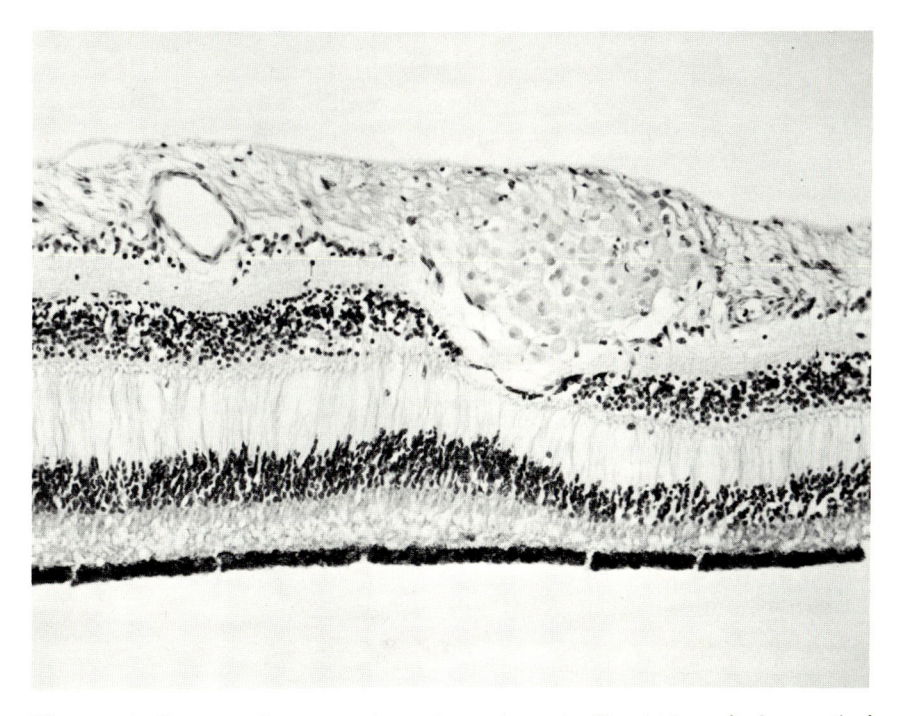

Figure 4.17. Cotton-wool spots such as those shown in Fig. 4.16 result from retinal ischemia and correspond to swollen axons in the nerve fiber layer. In this tissue section they superficially resemble cells. (H & E, × 130)

system, are extremely susceptible to hypoxia. In arterial occlusion, hemorrhage is not a feature, inasmuch as the blood is not under increased pressure and generally drains away.

Retinal Venous Disease

Venous obstruction interferes with the removal of metabolites. With retinal venous occlusion, the tissue receives blood from the arterial side and ischemia is not as pronounced as in arterial obstruction, and cotton-wool spots are generally not a feature. Venous occlusion is accompanied by an elevated intravascular pressure. This predisposes to retrograde flow through collaterals resulting in their dilation; hemorrhage from disrupted capillaries; and impaired absorption of interstitial fluid.

With central retinal vein occlusion, a hemorrhagic infarct of the retina develops. It is characterized by flame-shaped hemorrhages in the nerve fiber layer of the retina, especially around the optic disc, engorged tortuous veins, edema of the optic disc and retina, and occasionally cotton-wool spots. Although the vision is generally poor, it may recover surprisingly well considering the severity of the funduscopic appearance. The effect depends largely on the amount of damage produced by the accompanying hemorrhage which is cleared very slowly by macrophages. In contrast to arterial occlusion which usually develops suddenly, venous occlusion may develop slowly. The retina does not revert to normal; microaneurysms and cystoid degeneration around the macula may develop. An intractable closed angle glaucoma with severe pain and repeated hemorrhages commonly ensues 2–3 months after central retinal vein occlusion ("100 day glaucoma," "thrombotic glaucoma"). This often happens in eyes predisposed to or already having open angle glaucoma. It also results from the adhesions that develop from the peripheral anterior synechiae secondary to rubeosis iridis. Occasionally, there is recanalization of the occluded vein or the establishment of a collateral circulation. Central retinal vein occlusion is frequently accompained by arteriosclerosis of the artery which shares a common adventitial sheath within the nerve.

Retinal Arterial Disease

Arterial obstruction may be limited to a branch of the central retinal artery and, in these instances, the occlusion is usually at a bifurcation point.

With retinal arterial occlusion, neuronal damage from ischemia results in a sudden onset of symptoms. Intracellular edema develops and is reflected funduscopically in the living patient by a pallor of the retina (Fig. 4.18). With central retinal artery occlusion, this tends to be most marked around the macula where the ganglion cell layer has its greatest concentration of cells. The absence of ganglion cells in the center of the macula permits the retinal pigment epithelium and the highly vascularized choroid beneath it to stand out in sharp contrast to the surrounding pale retina as a prominent red spot ("cherry red spot"). This appearance may occur as soon as 10 minutes after arterial occlusion but usually requires 2–3 hours. The lack of retinal circulation reduces the retinal arterioles to delicate red threads, which may be difficult to trace a short distance from the optic nerve disc. The intravascular erythrocytes become conglutinated into groups with rouleaux formation. On funduscopic examination, these appear as red segments interspersed by clear areas—an appearance that has been likened to "cattle trucks" or "box cars." The retinal arteries retain their diminished caliber to become white cords. Eventually, after about 1 week, the retinal circulation is restored presumably by cilioretinal anastomoses. Central retinal artery

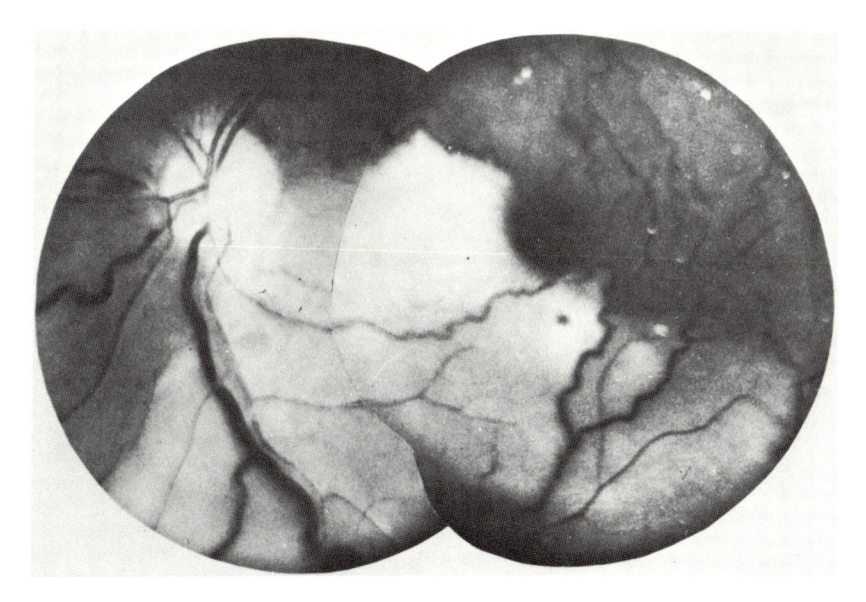

Figure 4.18. Occlusion of the inferior temporal branch of the retinal artery with segmental edema. The central macula retains its dark appearance.

occlusion may result in permanent blindness if the obstruction is of long duration. However, some recovery may occur if the ischemia lasts a relatively short time. Even in complete central retinal artery occlusion, a small temporal visual field usually remains due to anastomoses between the ciliary and retinal circulations. The loss of retinal ganglion cells and their axons, a sequel of retinal infarction, reflects itself as a shallow excavation of the optic nerve head and optic atrophy. Glaucoma following central retinal artery occlusion is uncommon.

SPECIFIC VASCULAR DISEASES

Atherosclerosis

Atherosclerosis, which is common in the hyperlipoproteinemias, may involve the ophthalmic, choroidal, and central retinal arteries. The central retinal arteries are affected especially where they penetrate the dural sheath of the optic nerve and at the lamina cribrosa. The intraocular arteries are seldom involved. The central retinal vein may become occluded due to endothelial proliferation beneath an arteriosclerotic central retinal artery. The affected vessels may become occluded by atheromatous material or thrombus.

Arteriolosclerosis

Arteriolosclerosis is characterized by thickening of the arteriolar walls with intimal hyalinization, medial hypertrophy, and endothelial hyperplasia. The retinal and choroidal vessels are commonly affected. Retinal arteriolosclerosis is evident in most individuals after the 5th decade. The thickened retinal arterioles become attenuated, increasingly tortuous, and of irregular caliber. At sites where veins and arteries cross each other, the veins may appear kinked, a phenomenon referred to as arteriovenous nicking. This appearance is not due to venous compression by a taut sclerotic artery, as the venous diameter does not vary on either side of the compression. At the site of the arteriovenous crossings, the two vessels share a common adventitial sheath, so that the abnormal arterial wall also involves the vein. Where the vascular crossings occur, the abnormal retinal arterioles appear funduscopically as parallel white lines that encase the arterioles (arterial sheathing). The reflecting properties of the sclerotic vessel wall result in alterations in the light reflex ("silver wiring"). Individuals with retinal arteriolosclerosis may develop superficial or deep retinal hemorrhages. These are usually small and in the nerve fiber layer. Severe involvement of the choroidal arteries is accompanied by secondary degenerative changes in the retinal pigment epithelium and outer retina.

Hypertension

Patients with long standing hypertension frequently have arteriolosclerosis. A necrotizing arteriolitis with fibrinoid necrosis and thrombosis of the precapillary retinal arterioles occurs in malignant hypertension. Hemorrhages and exudates, as well as ischemic changes such as cotton-wool spots and microaneurysms, develop in the retinal area supplied by occluded vessels. In patients with severe hypertension, the retinal arterioles are usually considerably narrowed and papilledema may ensue. In monkeys with experimentally produced hypertension, the stage of fibrinoid necrosis is preceded by extreme arteriolar vasoconstriction and the insudation of plasma into the degenerated vascular wall.

Giant Cell Arteritis (Cranial Arteritis, Temporal Arteritis)

Giant cell arteritis is characterized by inflamed arteries with numerous multinucleated giant cells. It usually occurs in individuals over the age of 65. It often involves the temporal arteries which develop prominent nodules and become painful. Occlusions of the short posterior ciliary arteries and less often of the central retinal artery and its branches occur. Involvement of the recurrent branch of the central retinal artery causes optic atrophy and blindness in many cases.

Aortic Arch Syndrome (Takayasu's Disease, Pulseless Disease)

A giant cell obliterative endarteritis of the aortic arch occurs particularly in women. The cause of the condition is unknown. Retinal neovascularization, microaneurysms, and blindness may follow retinal ischemia.

Arteriovenous Fistulae

A carotid cavernous fistula may follow traumatic severance of the carotid artery within the wall of the cavernous sinus; others result from rupture of an aneurysm of the carotid artery into the cavernous sinus. When this occurs, arterial blood enters the orbital veins, markedly increasing the intravascular pressure and causing congestion and edema of many ocular structures, including the optic disc and conjunctiva. Some cases develop a pulsating exophthalmos.

REFERENCES

Ashton, N. Neovascularization in ocular diseases. Trans. Ophthalmol. Soc. U.K. 81:145-161, 1961.

Ashton, N. Pathophysiology of retinal cotton-wool spots. Br. Med. Bull. 26:143-150, 1970.

Ashton, N., Dollery, C. T., Henkind, P., Hill, D. W., Patterson, J. W., Ramalho, S., and Shakib, M. Focal retinal ischemia. Br. J. Ophthalmol. 50:281-384, 1966.

Coats, G. Forms of retinal disease with massive exudation. R. Lond. Ophthalmol. Hosp. Rep. 17:440-525, 1907-1908.

Cullen, J. F. Occult temporal arteritis. Br. J. Ophthalmol. 51:513-525, 1967.

David, N. J., Klintworth, G. K., Friedberg, S. J., and Dillon, M. Fatal atheromatous cerebral embolism associated with bright plaques in the retinal arterioles. Neurology 13:708-713, 1963.

Fromer, C. H., and Klintworth, G. K. An evaluation of the role of leukocytes in the pathogenesis of experimentally induced corneal vascularization. I. Comparison of experimental models of corneal vascularization. Am. J. Pathol. 79:537-554, 1975.

Fujino, T., Curtin, V. T., and Norton, W. D. Experimental central retinal vein occlusion. Arch. Ophthalmol. 81:395-406, 1969.

Garner, A., Ashton, N., Tripathi, R., Kohner, E. M., Bulpitt, C. J., and Dollery, C. T. Pathogenesis of hypertensive retinopathy: an experimental study in the monkey. Br. J. Ophthalmol. 59:3-44, 1975.

Hayreh, S. S. Occlusion of the central retinal vessels. Br. J. Ophthalmol. 49:636-645, 1965.

Hayreh, S. S. Pathogenesis of oedema of the optic disc. Doc. Ophthalmol. 24:289-411, 1968.

Henkind, P., Charles, N. C., and Pearson, J. Histopathology of ischemic optic neuropathy. Am. J. Ophthalmol. 69:78-90, 1970.

Klintworth, G. K. The hamster cheek pouch: An experimental model of corneal vascularization. Am. J. Pathol. 73:691-710, 1973.

Manschot, W. A., and DeBruijn, W. C. Coats's disease: definition and pathogenesis. Br. J. Ophthalmol. 51:145-157, 1967.

Reinecke, R. D., and Kuwabara, T. Temporal arteritis. Arch. Ophthalmol. 82:446-453, 1969.

Schulze, R. R. Rubeosis iridis. Am. J. Ophthalmol. 63:487-495, 1967.

Wise, G. N., Dollery, C. T., and Henkind, P. The Retinal Circulation. Harper & Row, Publishers, New York, 1971.

5

DEVELOPMENTAL ANOMALIES

In the young human embryo (about 3-mm stage), the optic vesicles normally appear as bilateral diverticula of the prosencephalon. Should the evagination of this ocular anlage be defective, the eye may fail to form (*primary anophthalmia*). Anophthalmia may also follow degeneration of a previously formed optic vesicle (*secondary anophthalmia*). Instead of forming two separate optic vesicles, the optic plates of the embryo may give rise to a single medial optic vesicle (*cyclops*) or two fused vesicles (*synophthalmia*). True cyclopia is extremely rare and most instances of what appears externally to be a single eye represents a variable degree of ocular fusion. In both cyclops and synophthalmia, the ocular abnormality ranges in severity from one that is almost normal to a tuft of neuroectoderm surrounded by undifferentiated mesoderm. Moreover, the brain is abnormal with the telencephalon frequently not divided into two hemispheres. The frontonasal processes that normally develop into a nose form a single proboscis superior to the medial eye. Other malformations of the nose and mouth are often associated.

A degeneration of the neural plate in the region of the brain after the formation of the optic primordia results in *anencephaly*. When this occurs, the eyes continue to develop, and their external appearance is usually unremarkable, but the ganglion cell and nerve fiber layers of the retina are markedly hypoplastic. The sensory retina overrides the hypoplastic optic disc. Vascular endothelial hyperplasia within the eye is sometimes pronounced and may resemble the retinopathy of prematurity associated with oxygen administration.

The optic vesicles become invaginated during the 4th week (4–15-mm stage) to form a double-layered optic cup. The external layer forms the pigment epithelium of the retina, ciliary body, and iris. The inner layer forms the sensory retina, the nonpigmented epithelium of the ciliary body, the pigment epithelium of the iris, and the constrictor and dilator pupillae. Should the optic vesicle not invaginate, a *congenital cystic eye* or a nonattachment of the sensory retina to the retinal pigment epithelium (*congenital sensory retinal detachment*) ensues. Variable sizes of unilocular or multilocular orbital cysts are associated with a congenital cystic eye. A small eye may be associated with an orbital cyst (*microphthalmos with orbital cysts*) (Figs. 5.1, 5.2). The two adjacent epithelial layers of the ciliary body and iris may separate and line a cyst particularly at the pupillary margin and the anterior ciliary body. Such cysts are frequently bilateral and occur especially inferiorly and temporally. Iris cysts vary in size, the smaller ones are about 1 mm in diameter, whereas larger ones may obstruct the chamber angle and cause glaucoma or appear as dark, round masses displacing the iris forward. They are sometimes mistaken clinically for melanomas.

The choroidal fissure lies in the inferior portion of the optic vesicle while the embryo is about 4.5–22 mm long. Incomplete closure of all or part of it results in a *coloboma* which may extend anywhere from the optic nerve to the iris (Fig. 5.3). Both eyes are usually affected and the inferior nasal sector is typically involved. The coloboma seems white because the retinal pigment epithelium and uveal melanophores are absent at the colobomatous site. The overlying sensory retina is abnormal. When the ciliary body is involved, a notch appears in the crystalline lens from its inherent elasticity not being counteracted by zonular fibers. Although often sporadic, some colobomata are inherited as an autosomal dominant trait; others are due to a trisomy 13 or to thalidomide. The unfused margins of the bilayered optic cup may herniate through a gap in the choroidal fissure and form a cyst (*colobomatous cyst*), which can enlarge and totally replace the eye. Atypical colobomata of unknown cause and pathogenesis occur at the macula or at the optic nerve head. Atypical colobomata of the optic nerve (*optic pits*) are frequently single and involve the inferior temporal margin of the disc. About 30% develop a serous detachment of the macula.

The entire globe is sometimes smaller than normal (*microphthalmos*).

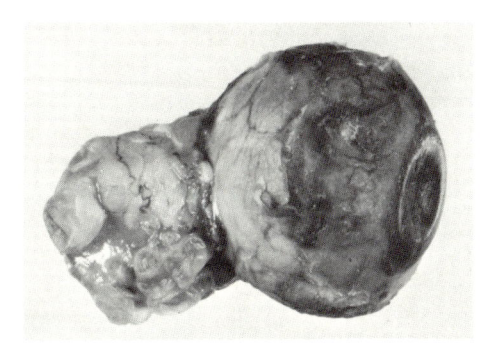

Figure 5.1. The ocular developmental anomaly known as microphthalmos with orbital cyst is characterized by a small malformed eye and a retrobulbar cystic mass.

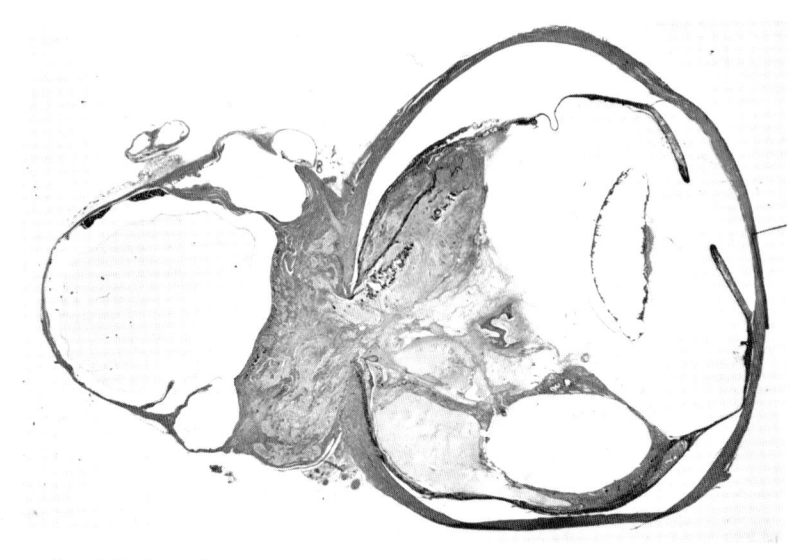

Figure 5.2. A horizontal tissue section of the eye shown in Fig. 5.1 demonstrates the cystic nature of the retrobulbar mass and a malformed retina. (H & E, × 3.2).

When extremely small, other ocular anomalies, such as aphakia, anterior segment aplasia, absence of pupil, and intraocular cartilage frequently coexist, as in individuals with trisomy 13.

HAMARTOMAS AND CHORISTOMAS

Non-neoplastic masses of malformed tissue can include tissue natural to the part (hamartoma) or may involve tissue elements that are not normally found in the involved site (choristoma). Some hamartomas and choristomas undergo neoplastic transformation, others often enlarge even though they are not neoplasms. Hamartomas are exemplified by hemangiomas, lymphangiomas, and nevi. The term phakomatosis has been used to include a variety of conditions which have in common hamartomas of the eye, skin, and brain; for instance, von Recklinghausen's neurofibromatosis, Sturge-Weber syndrome, tuberous sclerosis, and the von Hippel-Lindau syndrome. A teratoma is an example of a choristoma.

Hemangiomas

A variety of hemangiomas occur in and around the eye. Most appear at birth or during infancy. Some are composed entirely of haphazardly arranged capillaries (capillary hemangioma). A special type of capillary hemangioma found in infants is characterized by a prominent endothelial proliferation (benign hemangioendothelioma, infantile hemangioendothelioma, angioblastic hemangioma). Single or multiple lesions may occur in the orbit and retina. Thick or thin walled vascular channels may be widely dilated with some interspersing fibrous tissue. Such cavernous hemangiomas most often occur in the orbit and choroid, and rarely in the retina. Those in the orbit are usually encapsulated, soft, and lobulated, and comprise one of the commonest orbital masses. They can enlarge due to thrombosis of the vascular channels. Cyst formation and calcification may occur. Particularly capillary hemangiomas regress spontaneously with progressive fibrosis. Hemangiomas may bleed and be associated with hemosiderin-laden macrophages. Capillary hemangiomas often occur on the eyelids.

Hemangiomas are a component of several syndromes including von Hippel-Lindau, Sturge-Weber, and Maffucci. Retinal angioblastic hemangiomas are a component of von Hippel-Lindau disease. A cutaneous hemangioma in the distribution of the trigeminal nerve may be associated with homolateral hemangiomas of the choroid (Fig. 5.4), meninges, and cerebral cortex (Sturge-Weber syndrome). Homolateral congenital glaucoma is also common in the Sturge-Weber syndrome. Maffucci syndrome is characterized by bone lesions and cutaneous and choroidal hemangiomas. Choroidal hemangiomas may be mistaken clinically for malignant melanoma, retinoblastoma, or metastatic neoplasms in the choroid.

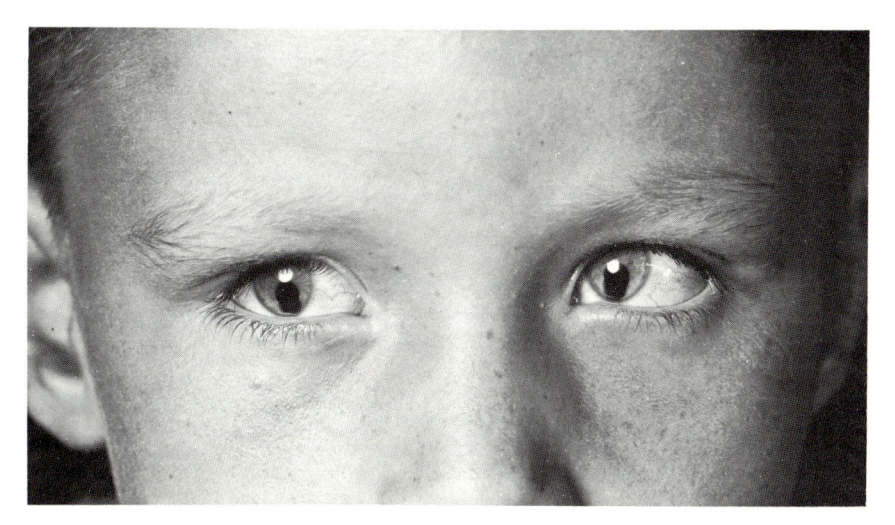

Figure 5.3. Bilateral colobomata of the iris. These are typically located downwards and inwards and may be associated with a defect of the retina, choroid, and ciliary body.

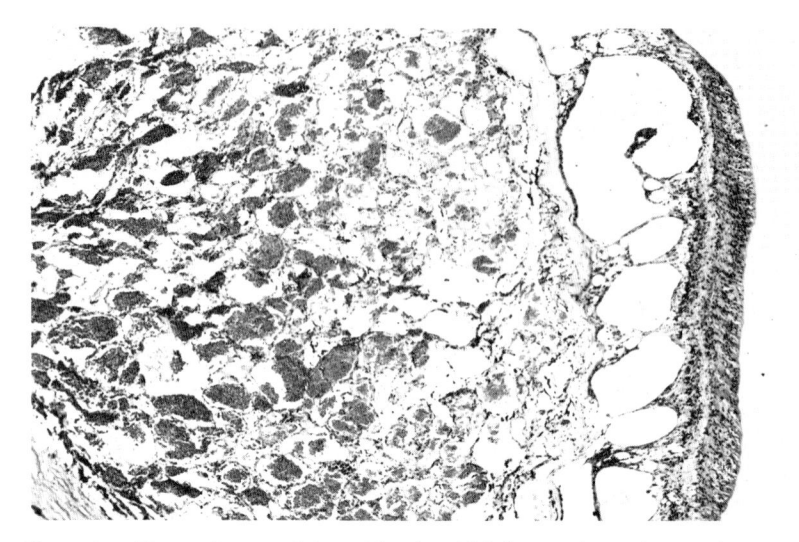

Figure 5.4. The retina overlying this choroidal hemangioma has undergone cystoid degeneration. (H & E, × 30).

Lymphangiomas

Angiomas composed of variable sizes of lymph channels occasionally occur in the conjunctiva, eyelids, and orbit. Hemorrhage into the lymphangiomatous spaces may cause the lesion to increase in size. A lymphocytic infiltration around the lymph channels is common. In adults, obstructions to conjunctival lymphatics may result in their dilation (lymphangiectasia).

Limbal "Dermoids" and "Dermolipomas"

Occasionally, part of the conjunctiva and/or cornea differentiates abnormally to form epidermis and hair follicles, as well as sebaceous and sweat glands. Such lesions (dermoids) are usually white or slightly yellow, well defined, round masses from which hairs often extend (Fig. 5.5). This anomaly occurs at the surface of an otherwise normal eye and often involves the inferior temporal limbus. Much adipose connective tissue may accompany the cutaneous elements (dermolipomas). Even ectopic lacrimal gland or brain may be evident in the lesion. Rarely, the entire cornea is replaced by fibrofatty tissue. Limbal dermoids exist alone or in association with mandibulofacial dysostosis or oculovertebral-auricular dysplasia (Goldenhar's syndrome). Components of Goldenhar's syndrome include unilateral or bilateral epibulbar dermolipomas (75% of cases), preauricular appendages, a unilateral or bilateral coloboma of the superior eyelid, tracheal fistula, and vertebral defects.

Dermoid Cysts

During the development of the face, the frontonasal and maxillary processes grow toward and fuse with each other. Rarely, the ectoderm becomes sequestered beneath the surface of the lines of fusion-producing cysts lined by stratified squamous epithelium with dermal appendages and connective tissue. These dermoid cysts contain desquamated hair, keratin, cholesterol, and sebum. Dermoid cysts appear in the skin of the forehead, especially in the supraorbital region and midline. A common site is the upper temporal orbital region where they appear as a painless swelling. Sometimes they occur nasally near the inner canthus. Dermoids rarely occur in the orbit behind the globe. Orbital dermoids are frequently attached to the periosteum and may excavate the underlying bone and extend into it. Some dumbbell-shaped orbital dermoid cysts extend partly extraorbitally. Dermoid cysts usually become apparent in early infancy, but occasionally not until late in childhood.

Miscellaneous Orbital Hamartomas and Choristomas

During the development of the brain, cerebral tissue and its meninges may herniate into the orbit as an orbital meningocele or encephalocele.

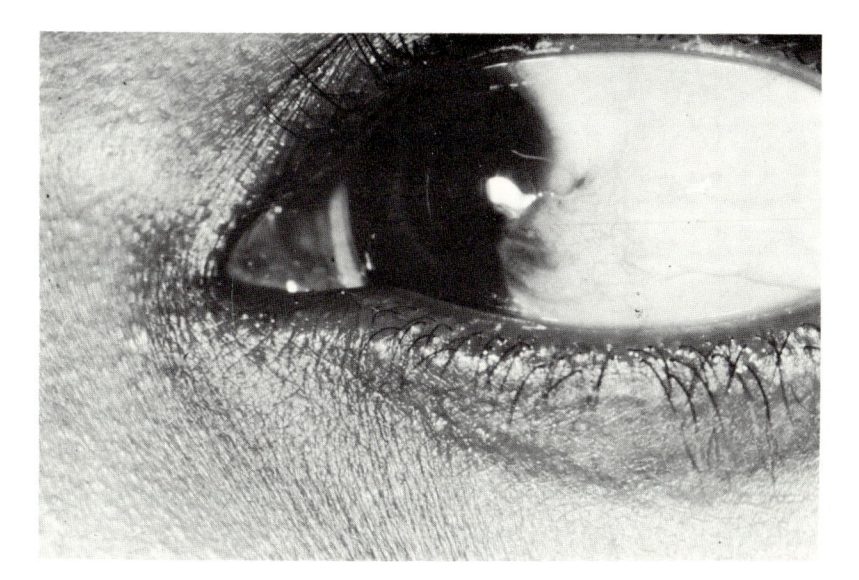

Figure 5.5. A limbal dermoid located inferior temporally, extending up onto the cornea.

Cerebral tissue that has completely or partly lost its connection with the brain can become sequestered in the orbit. Orbital teratomas composed of structures derived from ectoderm, entoderm, and mesoderm are uncommon. Some are cystic.

NEVI AND ANOMALIES OF PIGMENTATION

The term nevus has been applied to lesions present at birth, and to benign nests of cells, believed to be derived from epidermal dendritic melanocytic precursors. Nevus cells contain variable amounts of melanin pigment.

Cutaneous and Conjunctival Nevi

Nevi in the conjunctiva and skin are classified into junctional, intradermal (subepithelial), and compound types depending upon the location of the nevus cells. Junctional nevi occur most frequently in early life and consist of discrete nests of nevus cells within the cutaneous or conjunctival epithelium, and immediately beneath the epithelial basement membrane. With time, the cells within the epithelium are believed to migrate beneath the epithelium and to give rise to a compound nevus which has intraepithelial melanocytes and subepithelial nevus cells. Dermal (subepithelial) nevi represent a further stage of the evolution of this lesion. Here, all of the nevus cells reside beneath the epithelium. Deep nevi of the lids are often papillary and hairy. Conjunctival nevi most frequently occur near the corneoscleral limbus (Fig. 5.6) and inner canthus. Unlike those of the skin, they are often cystic due to inclusions of the surface epithelium. A lymphocytic infiltration is often present. Most pigmented nevi are present at birth or in early infancy. As a rule, the amount of pigment diminishes with the depth of the nevus.

Uveal Nevi

It is questionable whether the uveal tract contains nevus cells comparable to those of the skin and conjunctiva. However, focal aggregates of melanocytes commonly occur in the anterior stroma of the iris and choroid. They are rarely recognized in the ciliary body. When the accumulation of melanocytes is minimal, the term freckle is commonly applied, while larger discrete masses have been variably referred to as benign melanomas or nevi (Fig. 5.7). Sharply demarcated, roughly circular, flat, asymptomatic pigmented lesions commonly occur in the choroid, most frequently posteriorly. They are usually less than 0.5–4 disc diameters in size. Although probably present at birth, uveal freckles

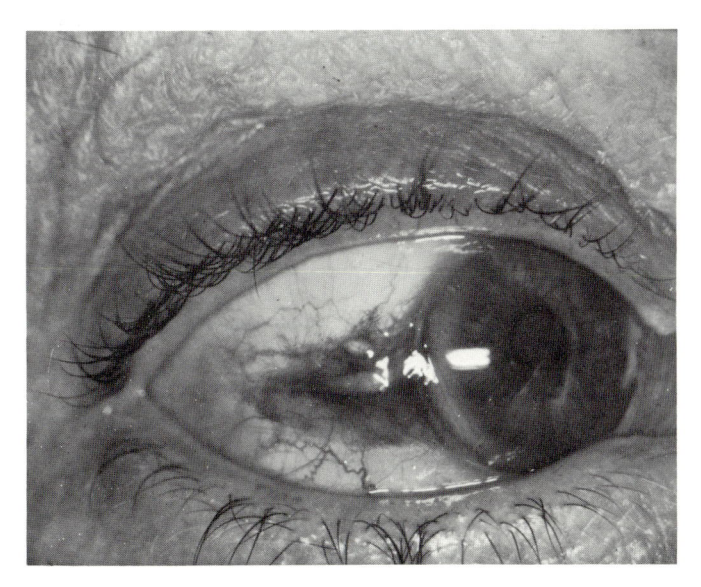

Figure 5.6. A conjunctival nevus showing a pigmented area with irregular edges located in the palpebral fissure adjacent to the limbus.

and nevi do not become evident until they contain sufficient pigment, usually at about puberty. With choroidal nevi, the overlying retina is normal. Most choroidal nevi remain unchanged for many years but some are thought to become malignant.

Melanocytoma

A melanocytoma is a special type of jet-black pigmented nevus that is most often located at the optic disc. In contrast to melanomas, it is common in the nonwhite races. Bleached tissue sections of this lesion reveal uniform, plump, polygonal cells with round nuclei.

Dermal Melanosis and Blue Nevi

Especially in the Mongolian race, but also in other races, a bluish discoloration may appear in the lumbosacral region (mongolian spot) and occasionally elsewhere (extrasacral mongolian spot). In such lesions, elongated, slender, stellate melanocytes are scattered among collagen bundles of the dermis. Melanocytes are present in the dermis of heavily pigmented individuals during a stage of embryonic development, and such lesions probably represent a delayed disappearance of these cells. The blue color results from light passing through a turbid skin before

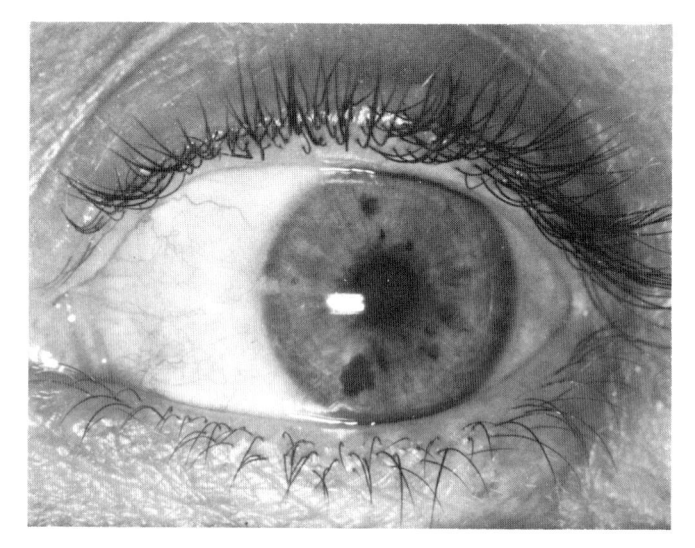

Figure 5.7. Iris nevi. Multiple flat, pigmented spots in the iris.

striking the melanin and having some of the light ray reflected. An aggregation of the pigmented melanocytes can form a benign melanocytic tumor (blue nevus). Such lesions are most frequently located deep in the connective tissue of the dermis. On rare occasions, they may be situated in the episclera or orbital tissue. Some blue nevi are more densely cellular and bizarre than the usual type (cellular blue nevus).

Ocular Melanosis

Numerous elongated dendritic melanocytes, like those in dermal melanosis, may be present in the conjunctiva, sclera, episclera, uvea, and orbit. The anomaly is sometimes clinically evident at birth. It can also appear during childhood or early adulthood, when it needs to be distinguished from other pigmented lesions. The condition is usually unilateral but is bilateral in about 5% of the cases. The melanocytes occur around perforating blood vessels and nerves in the episclera (episcleral melanosis) or throughout the sclera (scleral melanosis). Most of the globe can be involved, although the pigmentation is often confined to only a segment of the eye. The entire fundus may be black or chocolate colored and the ipsilateral iris is much darker than the opposite iris. Melanosis oculi may be associated with involvement of the ipsilateral eyelids and periocular skin of the face (the nevus of Ota, oculodermal melanocytosis). Other areas of the skin and mucous membranes and even the palate are sometimes involved. Melanosis oculi is more common in Orientals and less frequent in whites and blacks. Some individuals inherit melanosis oculi as a Mendelian dominant trait. A few cases of melanosis oculi have been associated with uveal melanomas, almost always in the affected eye.

ANOMALOUS DEVELOPMENT OF PARTS OF THE EYE
Cornea and Conjunctiva

With anomalous development, the cornea may be abnormally small (*microcornea*) or large (*megalocornea*). Diffuse or localized opacities occasionally develop in the cornea. The endothelium and Descemet's membrane are often absent in the affected region. An abnormal separation of the primary lens vesicle from the superficial ectoderm can result in a congenital central corneal opacity to which the iris is adherent. It is associated with an incompletely formed angle of the anterior chamber and often an anterior polar cataract (*mesodermal dysgenesis of the cornea, Peter's anomaly*). Rarely, the peripheral or entire cornea possesses the attributes of sclera (sclerocornea).

Sclera

In certain conditions, and most notably in osteogenesis imperfecta, the sclera is sufficiently transparent to permit the underlying choroid to appear a light blue ("*blue sclera*"). Nerves sometimes form a loop within the sclera, especially a few millimeters behind the limbus. When accompanied by numerous melanophores, an intrascleral nerve loop needs to be distinguished clinically from a transcleral extension of a melanoma. The sclera of many animals normally contains cartilage; occasionally cartilage forms in the human sclera.

Small cysts of the sclera occasionally occur in very young infants. They are lined by one or more layers of squamous epithelium, and sometimes communicate with the anterior chamber.

Lens

The somatic ectoderm that overlies the optic vesicle of the young embryo differentiates into the lens vesicle whose anterior cells form a single layer of cuboidal capsular epithelial cells. The remaining cellular elements develop into elongated primary lens fibers which eventually obliterate the cavity of the lens vesicle. The lens continues to enlarge throughout the life of the individual by cell divisions that take place at the equator of the lens. Newly formed cells elongate and form secondary lens fibers, which gradually lose their nuclei, mitochondria, and other cytoplasmic organelles. The fibers cover previously formed ones like the annual rings of a tree trunk. Successively formed strata produce zones which are readily visualized in the living patient with the slit lamp. From the center of the lens, these are referred to as embryonic nucleus, fetal nucleus, infantile nucleus, adult nucleus, and the cortex. Sutures form along the sites of apposition between secondary lens fibers. The appearance of the sutures varies with the stage of development. Toward the end of the 3rd month, they are shaped like an upright Y in front and an inverted Y behind (Fig. 5.8).

Congenital cataracts result from intrauterine rubella as well as from a wide variety of inherited and idiopathic conditions. The abnormalities of the lens are frequently nonspecific. In some congenital cataracts, there is a retention of the nucleated lens fibers. A subcapsular fibrous plaque may be present beneath the anterior lens capsule. Posterior polar cataracts are often associated with persistent remnants of the hyaloid artery (Fig. 5.9).

Other developmental anomalies of the lens include abnormalities in size, shape, and/or position. The lens may be small with an increased anterior and posterior curvature (*spherophakia*). When this occurs, the zonular fibers are easily seen through a dilated pupil and the lens is often displaced from its normal position. Spherophakia is a component of the Weill-Marchesani's syndrome.

Iris

Normally, the neuroectoderm and mesoderm grow in front of the lens to form the iris. If this does not occur, *aniridia* or hypoplasia of the iris results. The fringe of rudimentary iris frequently blocks the angle of the

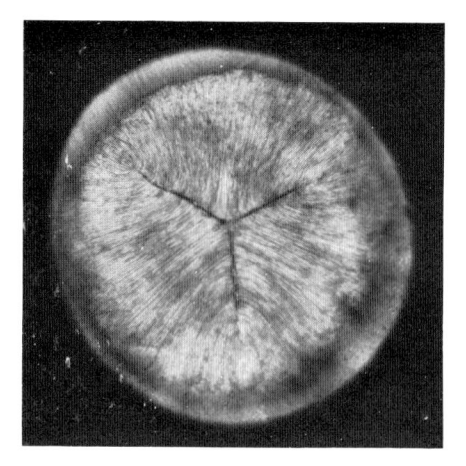

Figure 5.8. The triradiate suture is evident in this crystalline lens from the eye of a 3 year old child.

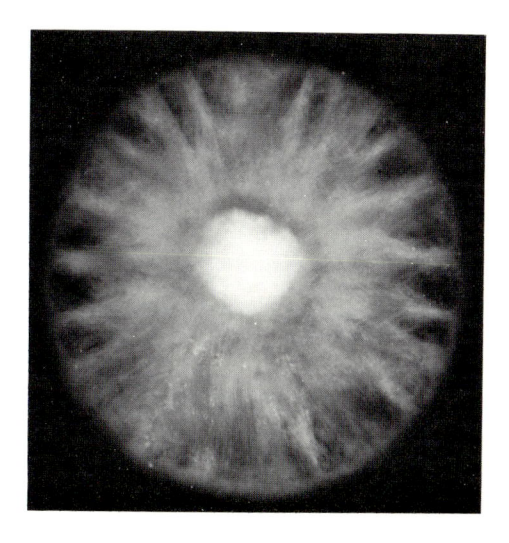

Figure 5.9. Posterior surface of the crystalline lens of a child with an inherited posterior polar cataract.

anterior chamber causing glaucoma. Aniridia may be inherited as an autosomal dominant trait. A small percentage of patients with bilateral sporadic aniridia have other associated ocular abnormalities (cataract, microphthalmia, posterior lenticonus, and capsule warts), microcephaly and mental retardation, genitourinary anomalies (horseshoe kidney, cryptorchism, hypospadias, pseudohermaphroditism), and an increased incidence of Wilms' tumor or other neoplasms by the age of 3 years (Brusa-Torricelli-Miller syndrome). Most individuals with this syndrome have been males.

Several other anomalies of the iris are also recognized. The pupils may be in an abnormal location (*corectopia*), unequal in size (*anisocoria*), or slit-shaped. Holes not surrounded by sphincters may be present (*pseudopolycoria*). Anomalies of iridial muscle or pigment may also occur.

Mesenchymal tissue extends from the iris across the pupil during the development of the eye. This pupillary membrane may persist, but usually regresses.

Anterior Chamber

Mesodermal tissue occupies the angle of the anterior chamber during embryonic and early fetal life. Normally, a recess forms between the cornea and iris at about the 6th month of intrauterine development. The ciliary muscles ordinarily originate from the scleral spur and adjacent connective tissue. Persistence of a fetal angle and an anomalous origin of the ciliary muscles from the trabecular meshwork, as well as other abnormal angles, are associated with a high incidence of glaucoma.

At the extreme periphery of Descemet's membrane, a collagenous ring (*Schwalbe's ring*) is often accentuated in otherwise normal individuals (*posterior embryotoxon*). Pectinate strands may extend from the anterior surface of a hypoplastic iris across the angle of the anterior chamber to the prominent Schwalbe's ring (*Axenfeld's anomaly*). A more advanced phase of this anomaly includes a failure of the peripheral iris to cleave from the cornea (*anterior segment cleavage syndrome*) and associated abnormalities of the teeth and bone (*Reiger's anomaly*). Malformations of the anterior chamber are often accompanied by defective zonules and a consequential subluxation of the lens.

Vitreous

During the 1st month of intrauterine life, the primary vitreous forms. The hyaloid artery soon enters it to extend from the optic nerve head to the lens where it gives rise to a vascular tunic around the lens (Fig. 5.10). With further development, the primary vitreous becomes less homogeneous. An anomaly known as *persistent hyperplastic primary vitreous* may result from the failure of the ectodermal and mesodermal elements of the primary vitreous to regress. This anomaly is characterized by a retrolental mass containing persistent portions of the hyaloid artery and vascular tunic of the lens. Frequently, islands of adipose connective tissue,

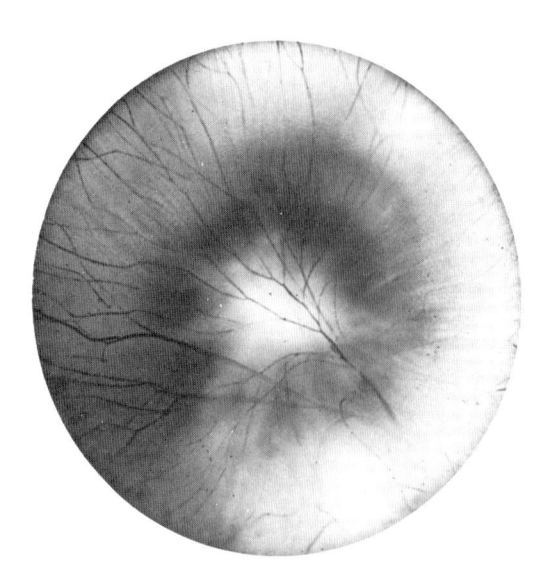

Figure 5.10. Delicate branches of the hyaloid artery are shown coursing over the posterior surface of the lens as they form a vascular tunic.

cartilage, undifferentiated neuroepithelium, and calcification are present. Elongated ciliary processes are attached to the lens or to the circumlental mass and drawn forward and inward. The lens is usually smaller than normal, distorted in shape, and cataractous at its posterior pole. The developing cataract is accompanied by swelling of the lens and shallowing of the anterior chamber. Iris bombé, glaucoma, and buphthalmos may develop as a result of the swollen lens or due to spontaneous hemorrhage into the vitreous or perilental area. The lens capsule may be ruptured by an ingrowth of fibrovascular tissue. The lens may become totally absorbed. As a rule, the condition is unilateral in an eye slightly smaller than normal. Leukocoria occurs in fully developed cases; the condition then needs to be distinguished clinically from a retinoblastoma. Traction of vitreoretinal adhesions occurs and may lead to detachment of the sensory retina. This course is usually progressive and at about 4 months of age, spontaneous hemorrhages commonly ensue. Persistent hyperplastic primary vitreous is frequently accompanied by other ocular developmental anomalies, such as ectopia lentis, corectopia, shallow anterior chambers, and colobomata of the iris and pupillary membranes.

In the 2nd month of life (13–17-mm stage), a secondary avascular vitreous (definitive vitreous) forms. The anterior fibers cling to the inner limiting membrane of the retina (vitreous base). It arches slightly around the macula and into the posterior lens capsule (hyaloideocapsular ligament). After the 40 mm stage, the hyaloid vessels within the primary vitreous normally atrophy, but remnants may persist. A short stub of a hyaloid artery may project into the vitreous at the optic disc, or a vestige may remain as a small opacity at the posterior lens capsule (*Mittendorf dot*). Eventually, the primary vitreous becomes compressed into the center of the globe where it occupies a narrow tubular area which extends anteroposteriorly from the lens to the optic disc (hyaloid canal, the canal of Cloquet). At the beginning of the 13th week, zonular fibers extend from the ciliary epithelium to the lens.

Retina

The outer layer of the optic cup leads to the retinal pigment epithelium, while the inner layer gives rise to the sensory retina. The potential space between these layers persists except at the site of exit of the nerve fiber layer of the retina at the optic disc and at the ora serrata. The sensory retina normally differentiates into several layers. Occasionally, aberrant retinal development results in a disorganized, irregular retina with immature visual receptors that form round or oval rosette-shaped structures. Frequently, an overgrowth of ciliary epithelium coexists. This anomaly, termed *retinal dysplasia*, often appears sporadically. It also occurs in the chromosomal anomaly trisomy 13 (Fig. 5.11). It can be produced in experimental animals with several viruses.

All layers of the retina except for the nerve fiber layer normally terminate at the edge of the optic disc. Sometimes the retinal pigment epithelium does not reach the disc margin and the choroid or sclera becomes visible through the retina. The retinal pigment epithelium may be heaped up and form a dark ring at the disc, usually on the temporal side.

Optic Nerve and Disc

After the differentiation of the retina, axons from the ganglion cells enter the optic stalk. During this phase of development, some retinal cells become sequestrated in the center of the optic disc and form a sheath around the hyaloid artery. Normally, this tissue atrophies with the hyaloid artery in later fetal life, but both may persist as an abnormal elevation of the disc (*Bergmeister's papilla*). Anomalous shapes of the optic nerve and disc sometimes develop. The normal, sharply demarcated margin of the optic disc may be blurred and a dirty gray color. Some anomalies of the optic disc need to be distinguished clinically from papilledema or inflammatory lesions of the optic disc. These include a congenital enlargement of the optic disc and calcified concretions in the optic disc, which may elevate the disc as well as project into the eye ("*drusen*") (Fig. 5.12).

The extension from the optic vesicle to the brain becomes constricted to form the optic nerve. The optic nerve becomes myelinated from the

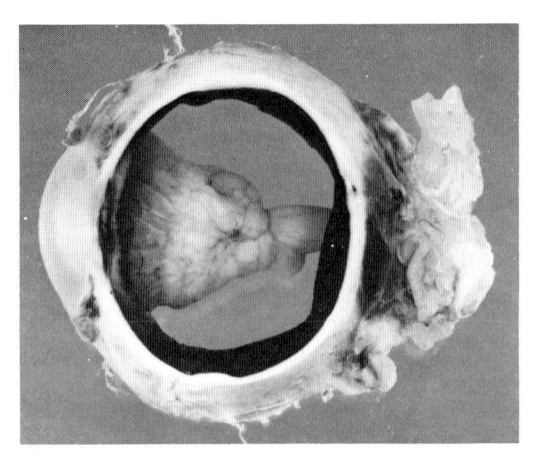

Figure 5.11. A retrolental mass of disorganized and malformed retina is present behind the lens in this eye of a child with a trisomy 13.

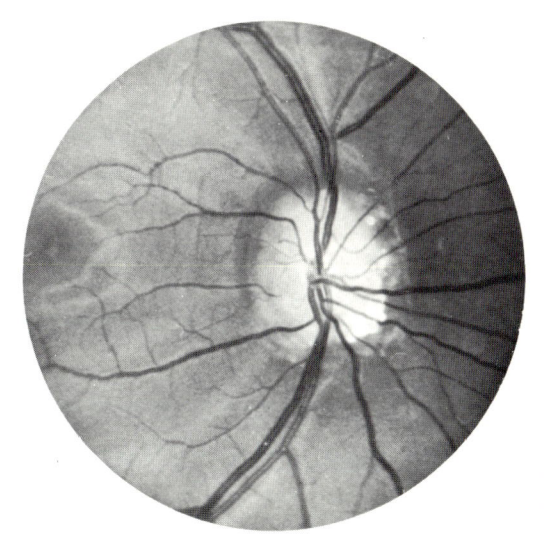

Figure 5.12. Drusen of the optic nerve head. Several nodules are present along the nasal margin of the optic disc.

chiasm towards the globe. Myelination usually ceases at the lamina cribrosa, but occasionally extends into the human retina producing a glistening white area usually continuous with the disc. When this occurs, oligodendroglia are located in the nerve fiber layer of the retina.

Eyelids

During the early development of the embryo (about 16-mm stage), a fold of mesoderm enclosed on both sides by a layer of epithelium covers the developing eye. It eventually becomes the eyelids, which remain fused until late fetal life. Occasionally, the eyelids separate imperfectly as a developmental anomaly (*ankyloblepharon*). The upper lid is formed from the medial and lateral frontonasal processes. Should these portions fail to unite adequately, a notch may result (*coloboma of the eyelid*). A semilunar vertical fold of skin conceals the medial canthus in early life in all races (*epicanthic fold*) (Fig. 5.13). Except in the Oriental races, it normally disappears before birth. It persists in Down's syndrome, thalassemia, Ehlers-Danlos syndrome, and as an inherited condition with an autosomal dominant mode of inheritance. Should the lid folds fail to form, the eye may not be evident from the surface (*cryptophthalmos*) (Fig. 5.14). This rare anomaly, which may be unilateral or bilateral, is frequently associated with other ocular and somatic congenital anomalies. Other anomalies of the eyelid include an absence of the eyelid (*ablepharon*), supernumerary rows of eyelashes (*distichiasis*), as well as anomalous lid folds, eyelashes, and eyebrows. In man, hair follicles rarely

84   The Eye

connect with the Meibomian glands although they commonly do so in several animals.

Blood Vessels

Aside from hemangiomas, which have already been discussed, other developmental anomalies of blood vessels include the persistence of embryonic or fetal vessels that normally regress with development; blood vessels that fail to form or that are aberrant in position or appearance; arteriovenous malfunctions, and, very rarely, aneurysms. Arteriovenous malformations of the retina and skin may be associated with angiomatous malformations of the brain, typically in the midbrain (*Wyburn-Mason syndrome*).

Lacrimal Glands and Passages

The lacrimal gland and/or passages occasionally fail to develop. Normally, the nasolacrimal duct opens into the inferior nasal meatus about the 3rd week after birth, but occasionally not until about the 6th month, and rarely not at all (*infantile stenosis of the nasolacrimal duct*). Lacrimal glands may develop in an abnormal location within the orbit as well as in the bulbar conjunctiva, cornea, and even within the eye. Ectopic orbital lacrimal tissue sometimes causes proptosis and may even

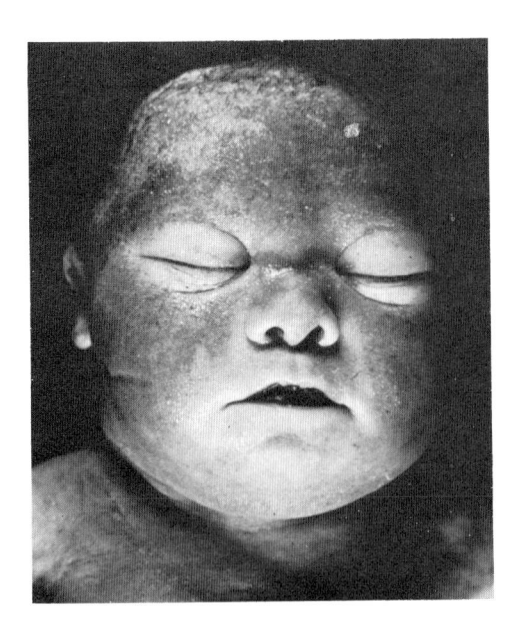

Figure 5.13. The medial canthi are covered by prominent epicanthic folds in the eyes of this infant with Down's syndrome.

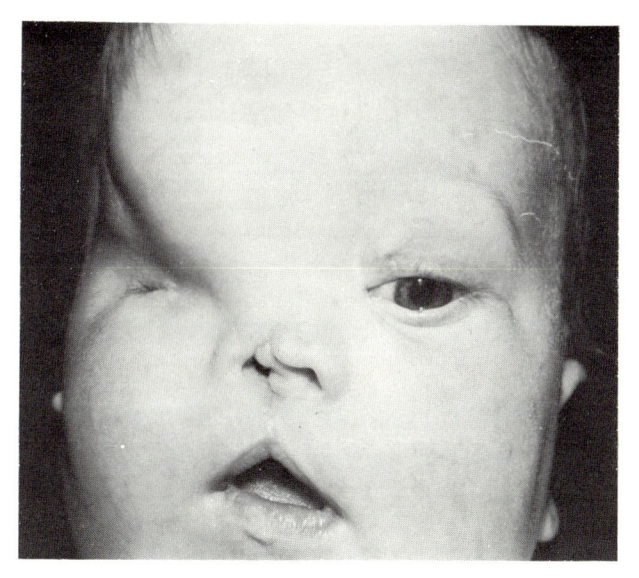

Figure 5.14. Partial cryptophthalmos. There is near absence of the lids of the right eye. A disorganized globe may be present in the right orbit. Other commonly associated deformities seen in this case include a cleft lip and palate, facial fissure with meningoencephalocele, and a coloboma of the upper lid of the opposite eye (surgically repaired).

become neoplastic. Muscle, nerves, cartilage, and/or dermal appendages occasionally surround ectopic lacrimal tissue.

Orbit

Osseous anomalies such as craniosynostosis, mandibulofacial dysostosis, and hypertelorism involve the orbit. Anomalous orbits are often associated with facial and other anomalies. The orbits may be widely separated. In craniosynostosis, one or more of the sutures of the skull may become prematurely closed, resulting in abnormally shaped orbits and heads of characteristic shapes: sagittal synostosis (*scaphocephaly*), all sutures (*oxycephaly*), coronal synostosis (*brachycephaly*), and portions of a suture (*plagiocephaly*). Premature fusion of the coronal sutures may be combined with hypoplasia of the maxilla (*Crouzon's disease, craniofacial dysostosis*). Orbital anomalies may be associated with the anterior segment cleavage syndrome (*Rieger's anomaly*).

SYNDROMES

Combinations of malformations commonly coexist. Because morphogenesis is a timely and sequential process, some developmental anomalies may be secondary to defects in one structure which compromised the development of subsequent structures. Noxious agents can have a deleterious effect on several tissues that are actively differentiating at the same time, and ocular malformations are commonly associated with independent developmental anomalies in other tissues. Other syndromes may be due to multiple primary defects, and yet, other syndromes may represent a coincidental association of unrelated anomalies.

Bilateral ocular abnormalities occur in association with systemic developmental anomalies in many different syndromes which result from the depletion or duplication of an extra chromosome. Several of these syndromes, including Down's and trisomy 13, were recognized before the chromosomal anomaly was established. In chromosomal anomalies, the

TABLE 5.1
Syndromes with Malformations of the Eye

Syndromes

A. **Inherited as autosomal dominant trait:**
   von Hippel-Lindau, von Recklinghausen's neurofibromatosis, Leri's pleonosteosis, Waardenburg-Klein, tuberous sclerosis (Bourneville's disease), Rieger, Franceschetti-Treacher Collins (mandibulofacial dysostosis), familial blepharophimosis, Crouzon (craniofacial dysostosis), Marfan, osteogenesis imperfecta, Pierre Robin, Alport

B. **Inherited as autosomal recessive trait:**
   Carpenter, Weill-Marchesani

C. **X-linked recessive:**
   Lowe

D. **Possible autosomal recessive inheritance:**
   Schwartz, Fraser

E. **Possible autosomal dominant inheritance:**
   Kenny

F. **Chromosomal abnormalities:**
   trisomy 21 (Down's syndrome, mongolism), XXXXY, trisomy 18 (Edwards' syndrome), trisomy 13 (Patau's syndrome), partial monosomy 18, chromosome No. 4 short arm deletion, chromosome No. 5 short arm deletion (cri-du-chat), chromosome No. 21 long arm deletion, coloboma of iris-anal atresia-extra chromosome syndrome, penta-X

G. **Viral:**
   rubella, cytomegalovirus

H. **Drugs:**
   thalidomide, lysergic acid diethylamide

I. **Idiopathic:**
   Sturge-Weber (encephalotrigeminal angiomatosis, encephalofacial angiomatosis), Riley, metaphysial dysostosis (Jansen type), Smith-Lemli-Opitz, Oculodentodigital, Goldenhar, Mieten, Cornelia de Lange, Rubinstein-Taybi, Hallerman-Streiff, Werner, Wyburn-Mason

resulting biochemical consequences must surely be of extraordinary complexity.

Several ocular abnormalities occur in Down's syndrome with greater frequency than in the general population: hypoplastic iris, keratoconus, widely separated eyes, and narrow palpebral fissures that run obliquely downward, as well as several types of cataract, epicanthic folds, and myopia. The iris is of a light blue color with white spots (Brushfield spots) scattered throughout. These tend to disappear if the iris becomes increasingly pigmented.

A wide range of ocular anomalies can occur in trisomy 13. The eyes tend to be smaller than normal. Anomalies related to the fetal fissure are a feature. Colobomata of the iris and ciliary body are usually present. Intraocular cartilage, retinal dysplasia, cataracts, an incompletely developed angle of the anterior chamber, and other anomalies are also common.

Aside from the aforementioned, numerous other ocular syndromes with developmental anomalies affect the eye (Table 5.1).

REFERENCES

ADDISON, D. J., FONT, R. L., AND MANSCHOT, W. A. Proliferative retinopathy in anencephalic babies. Am. J. Ophthalmol. 74:967–976, 1972.

ADELMANN, H. B. The problem of cyclopia. Q. Rev. Biol. 11:161–182, 284–304, 1936.

APPLE, D. J., AND BENNETT, T. O. Multiple systemic and ocular malformations associated with maternal LSD usage. Arch. Ophthalmol. 92:301–303, 1974.

APPLE, D. J., HOLDEN, J. D., AND STALLWORTH, B. Ocular pathology of Patau's syndrome with an unbalanced D/D translocation. Am. J. Ophthalmol. 70:383–391, 1970.

COULOMBRE, A. J. Cytology of developing eye. Int. Rev. Cytol. 11:161–194, 1961.

DUKE-ELDER, S., AND COOK, C. System of Ophthalmology, Vol. 3, Normal Abnormal Development, Part I: Embryology and Part II: Congenital Deformities. C. V. Mosby Co., St. Louis, 1963.

FERRY, A. P. Macular detachment associated with congenital pit of the optic nerve head. Arch. Ophthalmol. 70:346–357, 1963.

FONT, R. L., AND FERRY, A. P. The phakomatoses. Int. Ophthalmol. Clin. 12:1–50, 1972.

FRANCOIS, J. Congenital Cataracts, Charles C Thomas, Springfield, Ill., 1963.

FRAUMENI, J. F., AND GLASS, A. G. Wilms' tumor and congenital aniridia. J. Am. Med. Assoc. 206:825–828, 1968.

GINSBERG, J., AND BOVE, K. E. Ocular pathology of trisomy 13. Ann. Ophthalmol. 6:113–122, 1974.

HOEPNER, J., AND YANOFF, M. Ocular anomalies in trisomy 13-15. Am. J. Ophthalmol. 74:729–737, 1972.

HUNTER, W. S., AND ZIMMERMAN, L. E. Unilateral retinal dysplasia. Arch. Ophthmol. 74:23–30, 1965.

MANN, I. The Development of the Human Eye, 2nd edition. Grune and Stratton, Inc., New York, 1950.

MANN, I. Developmental Abnormalities of the Eye, 3rd edition. Grune and Stratton, Inc., New York, 1964.

MULLANEY, J. Ocular pathway in trisomy 18 (Edwards' syndrome). Am. J. Ophthalmol. 76:246–254, 1973.

REESE, A. B. Persistent hyperplastic primary vitreous. Am. J. Ophthalmol. 40:317–331, 1955.

SCHEIE, H. G., AND YANOFF, M. Peter's anomaly and total posterior coloboma of retinal pigment epithelium and choroid. Arch. Ophthalmol. 87:525–530, 1972.

SMITH, D. W. Recognizable Patterns of Human Malformation. W. B. Saunders Co., Philadelphia, 1970.

SUGAR, H. S. Congenital pits in the optic disc and their equivalents (congenital colobomas and coloboma-like excavations) associated with the submacular fluid. Am. J. Ophthalmol. 63:298–307, 1967.

Symposium on Surgical and Medical Management of Congenital Anomalies of the Eye. Transactions of the New Orleans Academy of Ophthalmology, C. V. Mosby Co., St. Louis, 1968.

YANOFF, M., RORKE, L. B., AND NIEDERER, B. S. Ocular and cerebral abnormalities in chromosome 18 deletion defect. Am. J. Ophthalmol. 70:391–402, 1970.

ZIMMERMAN, L. E. Melanocytes, melanocytic nevi and melanocytomas. Invest. Ophthalmol. 4:11–41, 1965.

ZIMMERMAN, L. E., AND FONT, R. L. Congenital malformations of the eyes. J. Am. Med. Assoc. 196:684–692, 1966.

6

NEOPLASMS AND CYSTS

A wide variety of neoplasms occur in the eye, eyelid, and orbit (Table 6.1). Like neoplasms elsewhere in the body, those of the ocular tissues do not arise from all cells with equal frequency. Most intraocular neoplasms arise from immature retinal neurons or uveal melanocytes. The retinal pigment epithelium commonly proliferates but seldom becomes neoplastic. Neoplasms rarely, if ever, arise from the epithelium of the lens in man.

Little is known about the cause of most human ocular neoplasms. The genetic constitution of the host is important in retinoblastomas and in neoplasms occurring in von Recklinghausen's disease of nerves (neurofibromas, schwannomas, and optic nerve gliomas). Actinic rays from the sun probably play a role in the genesis of at least some conjunctival carcinomas, as well as basal and squamous cell carcinomas of the eyelid. Like comparable carcinomas of the exposed skin, they are common in individuals exposed to excessive sunlight over many years and are usually associated with a connective tissue alteration attributable to solar irradiation (actinic elastosis). Races with marked cutaneous pigmentation rarely develop cutaneous carcinomas, possibly because of the protection offered by the skin pigment. On the other hand, albinos who lack melanin pigment are predisposed to cutaneous carcinomas. Squamous cell carcinoma often develops in the sun-exposed areas of the skin and conjunctiva of patients with xeroderma pigmentosum, an inherited disorder in which there is a deficiency of ultraviolet endonuclease, the enzyme that cleaves thymidine dimers. It has been noted particularly in the past that osteogenic sarcomas, chondrosarcomas, and other orbital sarcomas occur 10 years or longer after excessive local x-ray therapy for retinoblastoma or other neoplasms. Although there is no conclusive evidence of any human viral-induced neoplasm, some species develop ocular neoplasms in response to oncogenic viruses.

Ocular neoplasms destroy important structures and interfere with their function, sometimes resulting in blindness. Intraocular tumors may interfere with the drainage of aqueous humor and cause glaucoma. They

TABLE 6.1
Primary Neoplasms of Eye and Its Adnexa

I. Intraocular neoplasms:
Retinoblastoma
Medulloepithelioma
Teratoid medulloepithelioma
Glioneuroma
Adenomas and adenocarcinomas of ciliary epithelium
Melanoma
Leiomyoma
Neurofibroma
Schwannoma
Lymphoma

II. Neoplasms of eyelid and lacrimal drainage apparatus:
Benign

Keratoacanthoma
Papilloma of lacrimal sac
Leiomyoma
Inverted folliculoma (inverted follicular keratosis)
Calcifying epithelioma of Malherbe
Squamous papilloma
Seborrheic keratosis
Adenoma of sebaceous glands (Meibomian glands)
Adenoma of Krause's accessory lacrimal glands
Trichoepithelioma
Adenoma of sweat glands, and apocrine glands (Moll's glands)
Neurofibroma
Schwannoma

Malignant

Basal cell carcinoma
Squamous cell carcinoma
Malignant melanoma
Lymphoma
Adenocarcinoma of sweat gland
Meibomian gland carcinoma
Extramammary Paget's disease
Adenoacanthoma

III. Neoplasms of conjunctiva:
Dysplasia
Intraepithelial carcinoma (carcinoma in situ)
Squamous cell carcinoma
"Acquired melanosis"
Melanoma
Oncocytoma
Lymphoma
Neurofibroma

IV. Neoplasms of orbit:
A. *Neoplasms of lacrimal gland:*
   Mixed tumors
   Adenocystic carcinoma
   Mucoepidermoid carcinoma
   Oncocytoma

(TABLE 6.1—*continued*)

B. *Neoplasms of optic nerve:*
   Meningioma
   Optic nerve glioma
C. *Other orbital tumors:*

Benign

Neurofibroma
Schwannoma
Lipoma
Osteoma
Fibroma
Hemangioendothelioma
Hemangiopericytoma
Chondroma
Leiomyoma
Aneurysmal bone cyst
Fibrous dysplasia
Fibrous xanthoma
Myxoma

Malignant

Rhabdomyosarcoma
Malignant lymphomas
Plasma cell myeloma
Neurofibrosarcoma
Liposarcoma
Osteogenic sarcoma
Fibrosarcoma
Malignant hemangioendothelioma
Malignant hemangiopericytoma
Chondrosarcoma
Kaposi's sarcoma

may displace and distort the lens and make it cataractous. Exudates are common with choroidal melanomas and other intraocular neoplasms. Sensory retinal detachment may result from an intraocular neoplasm or an associated subretinal exudate. The retina overlying a neoplasm commonly degenerates due to pressure effects, compression of the choriocapillaris, and perhaps other factors.

Orbital tumors displace the eye forward causing proptosis (Fig. 6.1). Difficulties with extraocular movement may follow compression or infiltration of nerves or extraocular muscles.

Malignant neoplasms of the eye and ocular adnexa, like those elsewhere in the body, may invade blood vessels and become widely dispersed. Spread by lymphatics is limited to neoplasms of the eyelid and conjunctiva, as other ocular and orbital structures lack a lymphatic

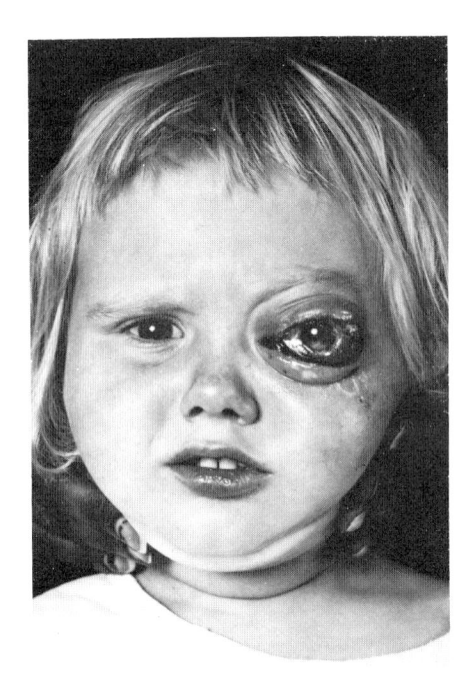

Figure 6.1. Proptosis of the left eye secondary to metastatic neuroblastoma to the orbit. The eye is displaced forward and laterally.

drainage. These lymphatics drain into the preauricular and cervical lymph nodes.

NEOPLASMS OF NEUROEPITHELIUM

Retinoblastoma

Retinoblastoma, the commonest intraocular malignant neoplasm in childhood, most frequently presents in the first year or two of life with a white pupillary reflex ("cat's eye reflex") (Fig. 6.2). Other presenting signs may be strabismus, poor vision, or a red, painful eye, often with secondary glaucoma. Retinoblastoma has been estimated to have an incidence of 1:20,000–34,000 per live births. It occurs with equal frequency in both sexes. Some retinoblastomas are inherited as a Mendelian dominant trait with incomplete penetrance, but most patients (about 95%) lack a positive family history. Even sporadic cases sometimes transmit the disease to their offspring as an autosomal dominant condition. There is a high incidence of retinoblastoma in individuals with a D-group depletion chromosomal abnormality. Bilateral neoplasms are common (approximately 30%). The offspring of survivors of retinoblastomas are especially prone to bilateral tumors.

Neoplasms and Cysts 93

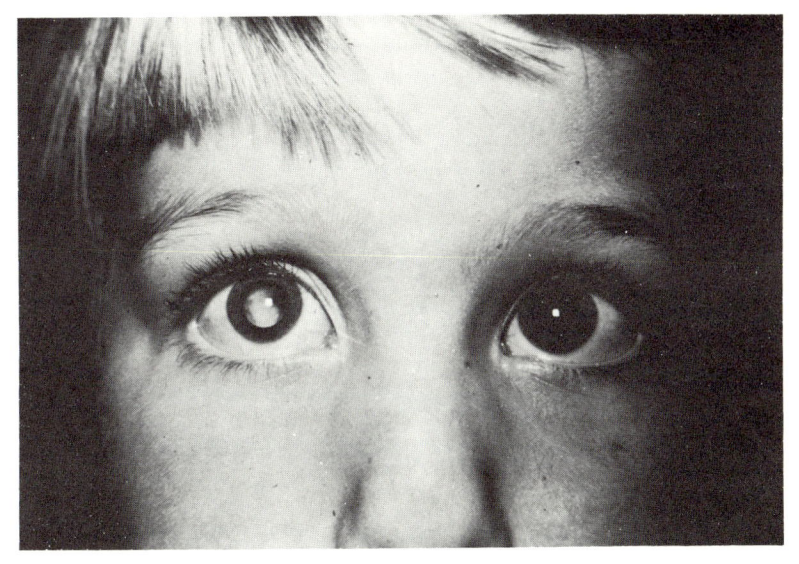

Figure 6.2. Leukocoria (cat's eye reflex). The white material filling the pupil of the right eye is a retinoblastoma.

The tumor has a pinkish white or creamy white color often with a scattered yellowish area of necrosis and white calcified flecks (Fig. 6.3). Some neoplasms extend behind the retina and displace it forward (exophytic retinoblastoma), others grow into the vitreous (endophytic retinoblastoma), while yet others are both endophytic and exophytic. Rarely is there any macroscopic tumor and minimal or any retinal detachment (diffuse infiltrating retinoblastoma).

Microscopically, the retinoblastoma is intensely cellular with a minimum of supporting stroma (Fig. 6.4). Several different morphologic patterns occur. Many retinoblastomas consist of densely packed round to oval neoplastic cells with hyperchromatic nuclei, scant cytoplasm, and abundant mitoses.

Tapered cells may arrange themselves radially around a central cavity demarcated by an inner limiting membrane (Flexner-Wintersteiner rosettes) (Fig. 6.5). This pattern represents differentiation towards visual receptors. In some retinoblastomas, the cellular arrangement resembles the fleur-de-lis (fleurette). In other instances, the neoplastic cells are distributed in no special arrangement. In common with many other neoplasms, the cells sometimes align themselves around blood vessels (pseudorosettes). Viable cells are generally arranged around the blood vessel while necrotic areas are very common further away. Necrosis in

94   *The Eye*

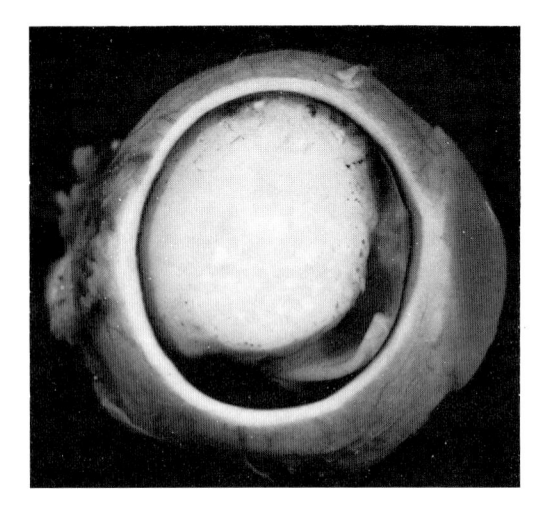

Figure 6.3. This eye is almost filled by a creamy colored intraocular retinoblastoma. Note the white flecks of calcification.

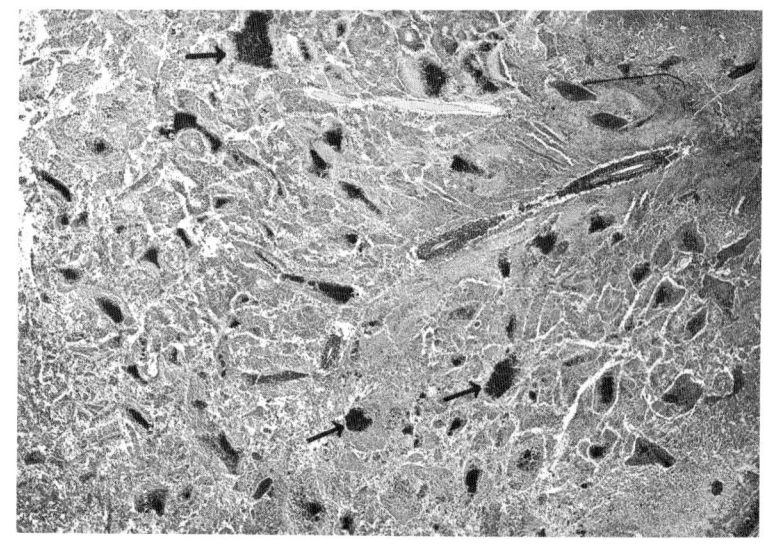

Figure 6.4 Intensely basophilic foci of calcification (arrows) are common in retinoblastomas. (H & E, × 9)

retinoblastomas rarely provokes a marked cellular response. Calcification is extremely common in areas of necrosis and may be evident on x-ray examination (Figs. 6.6, 6.7).

Multiple neoplastic foci often exist in the retina of the same eye. Some of these are thought to reflect a multicentric origin of the tumors; others reflect transplantations on the retina from neoplastic dissemination

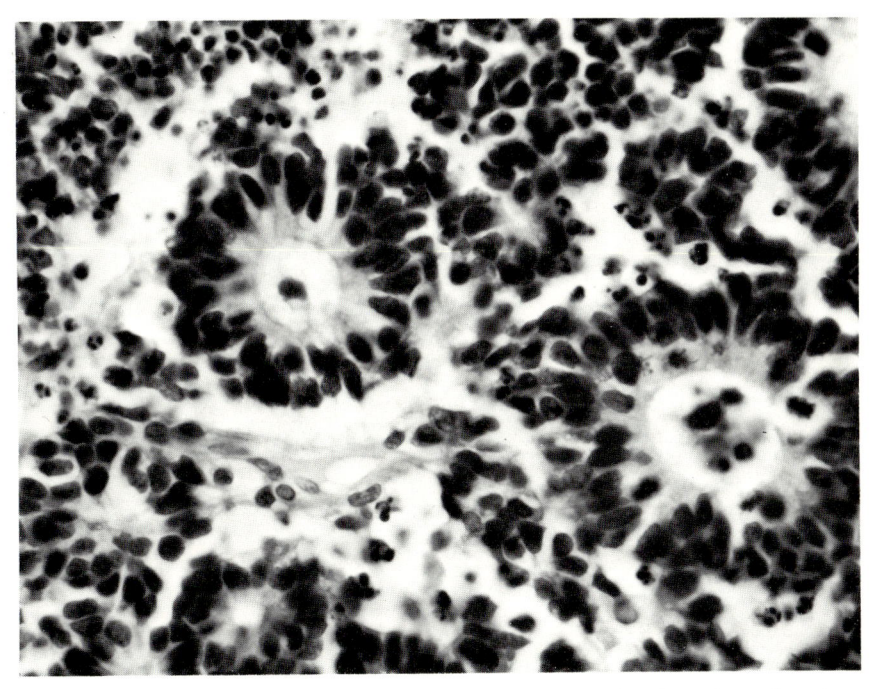

Figure 6.5. The neoplastic cells in retinoblastomas frequently align themselves into rosettes like these. (H & E, × 520)

through the vitreous. Retinoblastomas characteristically spread by direct extension along the optic nerve, by which route they may extend intracranially. Retinoblastomas frequently invade the highly vascular choroid. Blood-borne metastases often go to bone.

The overall mortality ranges from approximately 25% in the United States to near 100% in underdeveloped nations. With early diagnosis and modern therapy, 90% of patients with retinoblastomas survive. Some patients have survived retinoblastomas, only to die from apparently separate primary neoplasms, such as osteogenic sarcomas of the femur and Wilms' tumor. Retinoblastomas rarely will undergo a temporary or permanent spontaneous regression.

Medulloepitheliomas

The embryonal medulloepithelioma, an uncommon intraocular neoplasm, may be present at birth but does not usually become evident until 5 years of age. Unlike the retinoblastoma, it is not bilateral or multicentric. The lesion may produce early glaucoma. The microscopic

96   The Eye

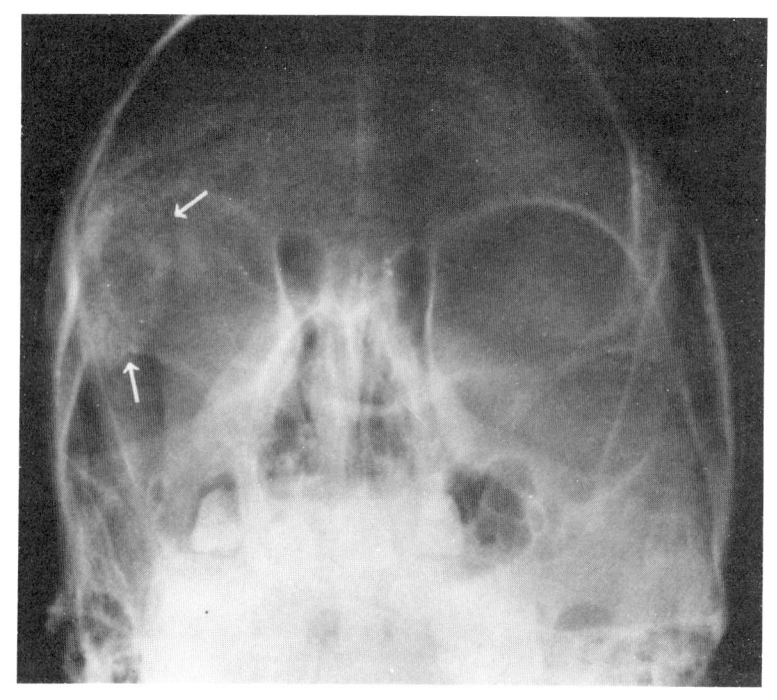

Figure 6.6. The radiodense area in the right orbit (arrows) was caused by calcification in a retinoblastoma.

morphology is variable. Elongated cells are aligned in multilayered rows or around an empty central cavity with an inner limiting membrane. The structure may resemble a net and at one time this lesion was referred to as a dikytoma (dikyton = net). Areas of the tumor may resemble the embryonic retina, pars ciliaris, retinoblastoma, and retinal pigment epithelium. An anaplastic variety is locally invasive.

A similar type of tumor may contain small islands of cartilage or striated muscle (teratoid medulloepithelioma). Such neoplasms may extend into the optic nerve and orbit. Medulloepitheliomas that occur in adults usually arise in an eye with severe inflammation. Metastases have been reported only rarely in this heterogeneous group of neoplasms.

Adenoma and Adenocarcinoma

An adenoma (benign epithelioma) of the epithelium of the ciliary body frequently occurs in the eyes of adults, especially in the elderly. It is usually less than 1 mm in diameter, unpigmented, and asymptomatic.

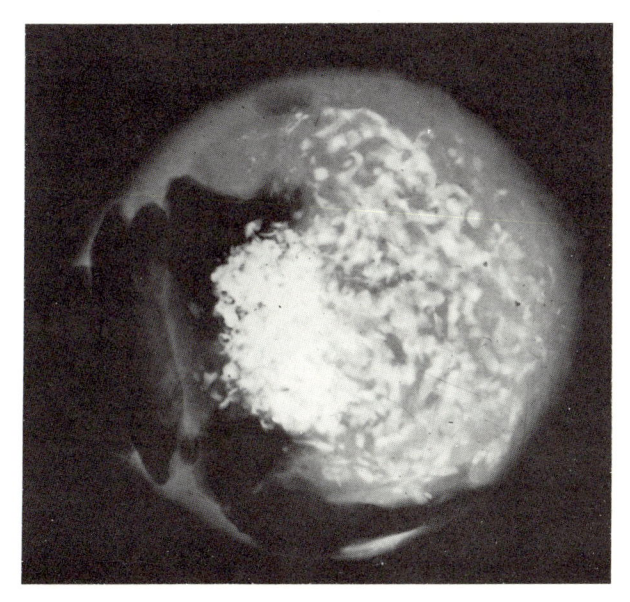

Figure 6.7. This roentgenogram of an enucleated globe with a retinoblastoma illustrates the characteristic speckled nature of the calcification.

Occasionally, it compresses the lens, causing cataracts and presses the iris forward. When it arises near the root of the iris, it is often partially pigmented. The pigmented cells often exhibit prominent vacuolization. Rarely does the tumor invade the ciliary body, angle, and iris (adenocarcinoma).

EPITHELIAL NEOPLASMS

Neoplasms of the Eyelids

Keratoacanthoma

This benign, self-healing, epithelial tumor is characterized by a central keratin-containing crater and a history of rapid growth. The lesion frequently resembles a horned toad scale. It may pose a diagnostic problem to the pathologist if it is biopsied and submitted for interpretation without an adequate clinical history because portions of the lesion may resemble a well differentiated squamous cell carcinoma microscopically.

Seborrheic Keratosis (Basal Cell Papilloma, Seborrheic Wart, Senile Wart)

This slightly elevated, well circumscribed, benign lesion is common on the eyelid and may resemble a basal cell carcinoma clinically. It can be light brown to black in color and does not extend into the dermis. Multiple lesions are frequently present. The crust is greasy and loosely attached.

Basal Cell Carcinoma

Basal cell carcinoma occurs much more frequently than squamous cell carcinoma (39:1). It often occurs on the lower lid and usually begins as a small, slight elevation (Fig. 6.8). It eventually evolves into a lesion of variable appearance. The tumor may possess pearly white margins and an excavated center (rodent ulcer). Nodular, scirrhous, pigmented, and erythematous types also occur. The gross appearance of the neoplasm reflects the various morphologic patterns that evolve. On microscopic examination, the lesion may be subdivided into cystic, pigmented, solid with squamous differentiation, scirrhous (with marked cutaneous connective tissue proliferation), and adenomatous (pseudoglandular) types. The biologic behavior of the tumor is independent of these morphologic

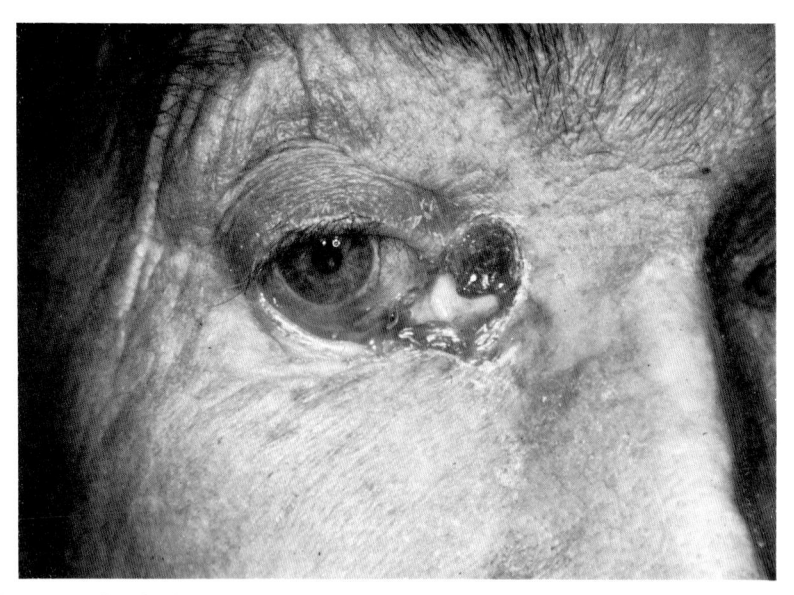

Figure 6.8. Basal cell carcinoma. Commonly seen in the medial canthal region, they rarely metastasize but may be locally invasive and recur after inadequate incision.

patterns. Basal cell carcinoma spreads by direct extension and those arising in the eyelid frequently invade the orbit particularly if inadequately treated. Distant metastases rarely occur.

Squamous Cell Carcinoma

Squamous cell carcinoma occurs predominantly in older individuals and often presents at the margin of the eyelid. Actinic keratosis (solar keratosis) and radiation dermatosis represent a phase in the evolution of the neoplasm. It begins as a wart-like keratic lesion. As the lesion enlarges, it frequently ulcerates and becomes fissured and indurated. Spread by lymphatics to the preauricular and submandibular lymph nodes can occur, but distant metastases are rare.

Other Epithelial Neoplasms

Adenocarcinoma of the Meibomian gland may simulate a Meibomian cyst or chalazion clinically. Sessile and pedunculated squamous papillomas are common at the margin of the eyelid. Cutaneous horns are often present near the margins of the inner canthus in middle-aged or elderly individuals. These lesions may occur on the surface of actinic keratosis, squamous cell carcinoma, or keratoacanthoma.

Epithelial Neoplasms of the Conjunctiva

Papillomas in the conjunctiva occur most commonly near the limbus. In young children, they may be multiple and recur after excision (recurrent juvenile conjunctival papillomatosis).

Dysplasia, intraepithelial carcinoma, and invasive squamous cell carcinoma can arise from the bulbar conjunctiva especially near the corneoscleral limbus on the temporal side of the globe (Fig. 6.9). Even when conjunctival carcinomas become large fungating exophytic masses, invasion of the globe, and metastases to the regional lymph nodes are rare.

Neoplasms of Lacrimal Gland

Mixed Tumor (Pleomorphic Adenoma) of Lacrimal Gland

The mixed tumor of the lacrimal gland, like its counterpart in the salivary gland, contains epithelial and mesenchymal elements believed to arise from embryonal cells. The epithelial cells are variable in shape and arranged in a variety of patterns. Epidermoid nests and keratin-containing cysts may occur. The prominent mucoid stroma may contain

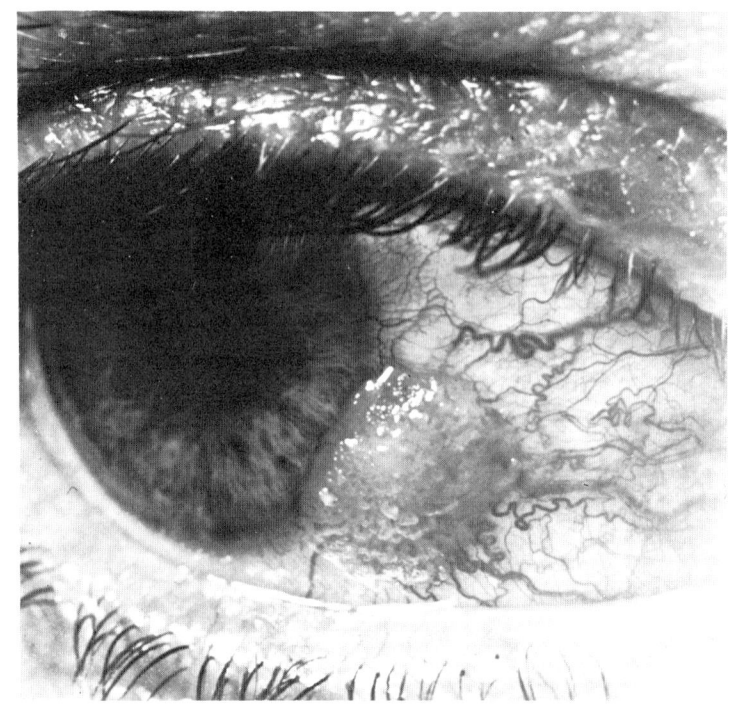

Figure 6.9. Squamous cell carcinoma of the conjunctiva located at the limbus.

cartilage and even bone. Although usually slow growing, the neoplasm is highly infiltrative and may invade the orbital bone. The neoplasm recurs after incomplete excision. Rarely, the morphologic features of adenocarcinoma or squamous cell carcinoma develop within the neoplasm (malignant mixed tumor). Such areas are friable and prone to hemorrhage and necrosis. Distant metastases from malignant mixed tumors can occur.

Adenoid Cystic Carcinoma (Adenocystic Carcinoma, Cylindroma)

The adenoid cystic carcinoma is the commonest malignant neoplasm of the lacrimal gland. This highly invasive neoplasm occurs relatively more frequently in the lacrimal then in the major salivary glands. Spread is by direct extension into the orbit, lids, subconjunctival tissue, and base of the brain. The tumor infiltrates contiguous bone and may produce hyperostosis which is detectable on x-ray examination. There is a marked tendency toward perineural spread which accounts for the excruciating pain that so commonly accompanies this neoplasm. Spread by lymphatics to the preauricular and cervical lymph nodes is late, as are blood-borne metastases.

Mucoepidermoid Carcinoma of the Lacrimal Gland

A mucoepidermoid carcinoma of the lacrimal gland containing mucus-secreting and epidermoid cells is rare.

Neoplasms of the Lacrimal Sac

Papillomas, squamous cell carcinomas, and mucus-secreting adenocarcinomas rarely arise in the lacrimal sac. They have many similarities to comparable neoplasms of the nose and paranasal sinuses.

NEOPLASMS OF MUSCLE

Rhabdomyosarcoma

Rhabdomyosarcoma is the most common primary malignant orbital neoplasm in children in the United States and several other countries. Pleomorphic, embryonal, well differentiated, and alveolar types are recognized. The embryonal type is the most common and is usually located in the upper inner aspect of the orbit. Cross-striations due to the filaments of actin and myosin are a prominent feature in the well differentiated type. In the other varieties, these cross-striations frequently are not evident in the primary orbital neoplasm, but may be identified in recurrences and metastases. The neoplasm is thought to be derived mainly from primitive multipotential cells. Although the neoplasm has been observed in all ages, it is rare in adults and most patients are under the age of 16. A few rare, well differentiated "rhabdomyosarcomas" have been reported in the iris but these have not metastasized.

Leiomyoma

Rarely do leiomyomas occur in the iris, ciliary body, orbit, and dermis. In the iris, most are near the pupil margin and arise from the sphincter pupillae. They may project into the angle or remain circumscribed. They are pink when vascularized and light brown when pigmented.

NEOPLASMS OF NERVES

Optic Nerve Glioma

Optic nerve gliomas account for 67% of optic nerve neoplasms and about 1–2% of orbital tumors. About 10% of reported cases of optic nerve glioma are associated with von Recklinghausen's disease of nerves. Most optic nerve gliomas are confined to the orbital portion of the nerve. Some infiltrate the nerve from the chiasm or adjacent hypothalamus. They are commonly unilateral but occasionally bilateral. Seventy-five percent occur during the 1st decade, usually under the age of 5. They occasionally develop in adults. They are prone to infiltrate the nerve sheath but do not penetrate the dura. They commonly invade the subarachnoid space and excite a meningeal fibromatosis.

The tumor causes a painless axial exophthalmos. Because the neoplasm destroys the myelinated axons in the optic nerve, visual loss and optic atrophy are early clinical abnormalities. Because the extraocular muscles are not infiltrated, limitation of extraocular movement is late and generally not marked. Optic nerve gliomas rarely present clinically with central retinal vein occlusion or glaucoma due to rubeosis iridis. The tumor may enlarge the optic foramen and provide a roentgenologic feature that is often of diagnostic importance. The vast majority of neoplastic cells are astrocytes, but oligodendroglia may contribute to the gliomatous mass. Rosenthal fibers and cystoid spaces with mucoid material sometimes occur. The neoplasm merges with the contiguous nerve and may transgress the lamina cribrosa to penetrate into the retina and choroid. Optic nerve gliomas are slow growing neoplasms that do not metastasize. Death may result from an intracranial extension or from independent associated intracranial neoplasms.

Meningiomas

Meningiomas are slow growing neoplasms which arise from the meninges. Orbital meningiomas occur over a wide age range including childhood. They are more common in females than males (5:1), and may be a manifestation of von Recklinghausen's disease of nerves. Most meningiomas are discrete globular masses, others consist of sheets of tumor cells ("en plaque"). The majority of orbital meningiomas arise from the sphenoidal ridge, olfactory groove, or suprasellar location within the cranium. En plaque meningiomas of the outer one-third of the sphenoidal ridge are particularly prone to invade the orbit. Other meningiomas arise from meninges of the optic nerve, especially near the optic foramen. Meningiomas compress and distort adjacent tissues. The pial sheath of the optic nerve is usually not penetrated. Sometimes the optic nerve, extraocular muscles, sclera, choroid, and retina are infiltrated. Meningiomas may invade the contiguous bone and provoke hyperostosis, which can be evident radiologically.

Orbital meningiomas displace the eye forward and cause visual impairment and optic atrophy due to pressure on the nerve. Intracranial meningiomas sometimes cause ipsilateral optic atrophy and contralateral papilledema (Foster Kennedy syndrome). Various histologic variants of meningioma are recognized but these bear no relationship to the biologic behavior of the tumor, and hence are of no practical importance. Even with highly invasive meningiomas, metastases are extremely rare.

Other Neoplasms of Nerves

Schwannomas (neurilemomas), neurofibromas, and neurofibrosarcomas arise from the sheath of peripheral and cranial nerves. They are uncommon in the eye, orbit, lids, and conjunctiva and are usually a manifestation of von Recklinghausen's disease of nerves. Schwannomas are usually solitary, slow growing, circumscribed, encapsulated, spherical, or fusiform neoplasms that compress the adjacent nerve. Neurofibromas are derived from, and composed of, both fibroblasts and Schwann cells. They are less well defined than schwannomas and merge with the involved nerve. The skin overlying neurofibromas is often thick and redundant and the underlying lesion often feels like "a bag of worms." Some neurofibromas become sarcomatous and exhibit cellular anaplasia, pleomorphism, numerous mitoses, and giant cells. Infiltration of adjacent tissue and metastases may occur with such neurofibrosarcomas.

LYMPHOMAS

For many years, the terminology of malignant lymphomas has been confusing and controversial. Currently, they are divided into Hodgkin's disease and other malignant lymphomas, and the most widely accepted classification is that of Rappaport. This is based upon the identification of cell types and their degree of differentiation, and the architectural patterns of proliferation (diffuse or nodular). Lymphomas with follicle-like structures are designated as nodular (or follicular), whereas those which lack them are termed diffuse. Each type may be differentiated into lymphocytic well differentiated, lymphocytic poorly differentiated, mixed cell (lymphocytic-histiocytic), histiocytic, and undifferentiated. The term nodular was introduced because of the confusion and controversy regarding whether the so-called follicles arose from or were related to the active secondary lymphoid nodules.

Diffuse Lymphocytic Lymphoma, Well Differentiated (Lymphocytoma, Lymphocytic Lymphoma)

Especially in middle and advancing age, lesions that consist almost exclusively of small lymphocytes with no mitoses and few or no follicles frequently occur in the orbit and conjunctiva. They commonly occur bilaterally. The lymphocytic infiltrate is occasionally circumcorneal. The rectus muscles are often involved in the orbit. This lymphomatous disorder may remain localized to the orbit and/or conjunctiva. However, a generalized lymphoproliferative disorder may be evident at the time of the orbital presentation or subsequently.

Diffuse Undifferentiated Lymphoma (Burkitt's Lymphoma)

In 1958, Burkitt drew attention to a peculiar malignant lymphoma that was prevalent in Uganda. Since then, the entity has been recognized in many parts of the world. The orbit is commonly involved in Burkitt's lymphoma which is the most frequent cause of unilateral proptosis in young children in areas where the condition is endemic. Even with extensive orbital involvement, the globe is rarely invaded. A herpes-like virus (Epstein-Barr virus) has been implicated as a possible causal agent. In endemic areas, the disease occurs in children, although immigrants to endemic areas may develop the disease later in life. The tumor contains fairly uniform cells (about twice the size of small lymphocytes) with relatively large nuclei and scant cytoplasm. Larger cells are dispersed throughout the neoplasm giving a "starry sky" appearance. These cells within the tumor are actively phagocytic. In the absence of therapy, the disease is generally fatal in 3–6 months, but with chemotherapy, 5-year survivals have been reported.

Other Lymphomas

A nodular lymphoma (follicular lymphoma, giant cell follicular lymphosarcoma) seldom presents first in the orbit, conjunctiva, or eyelids. Nodular lymphomas are extremely rare in blacks.

A histiocytic lymphoma (reticulum cell sarcoma) is uncommon in the orbit and appears grossly as a white, soft tumor consisting of large cells with vesicular nuclei and prominent nucleoli. On rare occasions it occurs within the eye and ocular symptoms may be the first manifestation of the disease.

Hodgkin's disease rarely involves the ocular and orbital tissues.

Leukemia

The intraocular structures and the orbit commonly are infiltrated by immature leukocytes in leukemia, especially as a terminal manifestation. A diffuse infiltration of the choroid by immature leukocytes is common. In myeloid leukemia, intravascular accumulations predominate, while in lymphatic leukemia, the neoplastic cells tend to be mainly in the choroidal stroma. When the iris is involved, hemorrhage into the anterior chamber may result. Retinal involvement may be associated with hemorrhages. Granulocytic sarcoma (granulocytic leukemia) commonly involves the orbit causing proptosis.

Multiple Myeloma

In multiple myeloma, flame-shaped hemorrhages, microaneurysms, and exudates may occur in the retina. Papilledema and central retinal vein thrombosis also occur. Neoplastic masses of plasma cell precursors occasionally are present in the orbit. On rare occasions, iridescent crystals deposit throughout the cornea and conjunctiva in multiple myeloma.

MELANOMAS

The term melanoma designates a neoplasm derived from melanocytes which possess the characteristic property of melanin formation, even though this function may be dormant or lost under certain conditions. It does not signify a pigmented tumor containing melanin. Some melanomas are amelanotic and all pigmented neoplasms are not melanomas. For instance, some basal cell carcinomas, squamous cell carcinomas, meningiomas, and leiomyomas are pigmented. Melanomas can arise from the melanocytes in the choroid, ciliary body, iris, conjunctiva, or eyelid.

Melanomas of the Uvea

Uveal melanomas comprise the most frequent, primary malignant intraocular neoplasm in the United States and several other countries (Fig. 6.10). They are rare in blacks. There is no sex predisposition. Most occur in the choroid, especially the posterior part. The majority of uveal melanomas occur after the age of 50; they are rare before puberty. Those in the iris tend to present clinically earlier than those in the choroid and ciliary body, perhaps because of their location. Almost all uveal melanomas are unilateral.

Malignant melanomas are usually circumscribed and often displace and invade Bruch's membrane causing a collar stud or mushroom-shaped mass (Fig. 6.11). Some melanomas spread diffusely in the choroid giving rise to a flat lesion which causes a gradual visual deterioration over many years or presents clinically as an extraocular mass, or

106   The Eye

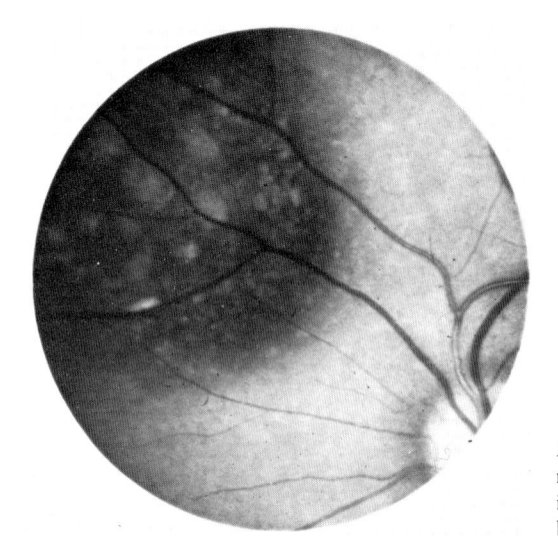

Figure 6.10. Malignant melanoma of the choroid. A dark mass is visible beneath the retinal blood vessels.

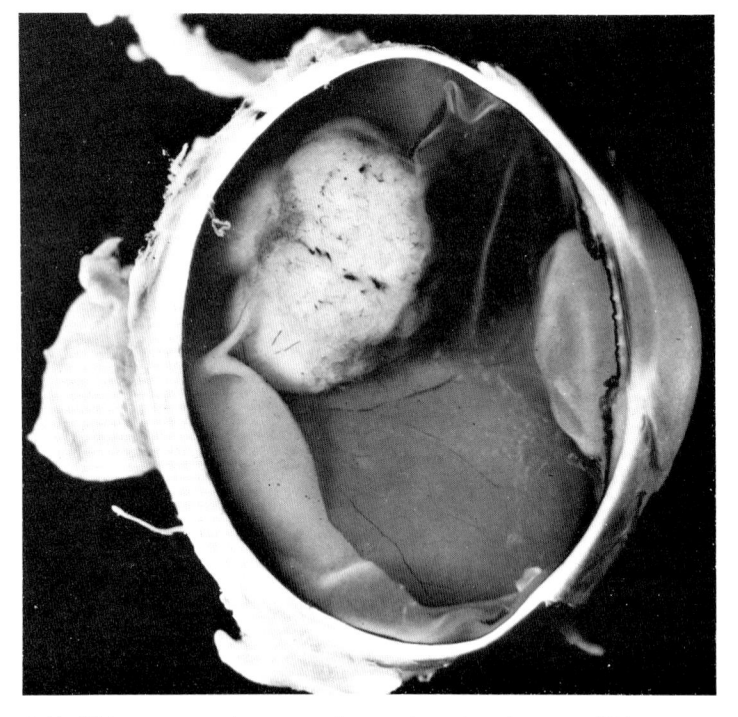

Figure 6.11. This eye contains a mushroom-shaped melanoma. This appearance is common with choroidal melanomas and results from the invasion of the neoplasm through Bruch's membrane.

with widespread metastases. Melanomas of the ciliary body and iris may extend circumferentially around the globe ("ring melanoma"). Melanomas adjacent to the optic disc are rare. A subretinal exudate, sensory retinal detachment, and degeneration of the overlying retina is common. Glaucoma is common and can result from one of several mechanisms. These include obstruction of the angle of the anterior chamber or trabecular meshwork by living or necrotic neoplastic cells, peripheral anterior synechiae, and anterior displacement of the iris against the filtration angle. The presenting clinical feature of some melanomas is intraocular hemorrhage, a cataract, glaucoma, or retinal detachment. Melanomas of the uvea are commonly vascular. Necrosis is sometimes present and may induce a marked inflammatory reaction including orbital cellulitis. Eyes enucleated with glaucoma, retinal detachment, intraocular hemorrhage, ocular inflammation, or phthisis bulbi sometimes contain clinically unsuspected melanomas. This underscores the importance of having all enucleated globes submitted for histologic examination.

Melanomas of the uvea vary in color. Most have some degree of pigmentation. Orange lipofuscin pigment is evident over the surface of some choroidal melanomas. As a rule primary choroidal melanomas do not have distant metastases at the time of their clinical presentation. Based on the microscopic appearance of uveal melanomas, they have been subdivided into spindle A, spindle B, fascicular, necrotic, mixed, and epithelioid types.

Uveal melanomas may spread transclerally into the orbit at points where blood vessels and nerves penetrate the sclera (Fig. 6.12). Unlike retinoblastoma, they rarely infiltrate the optic nerve, but may do so in the presence of glaucoma.

The prediction of the biologic behavior of uveal melanomas in a particular patient is difficult, if not impossible. A small one may widely disseminate and be fatal soon or many years after the primary is excised. However, long term follow-ups on a large series of patients with ocular melanomas have shown that the course of the patient after enucleation of eyes with melanomas can be correlated to some extent with the size, site, and morphology of the neoplasm. The larger the tumor, the more unfavorable the course; the greater the amount of reticulum and the smaller the lesion, the better the prognosis. Melanomas of the iris usually grow slowly, rarely metastasize, and have a much better prognosis than those in the choroid and ciliary body.

Several choroidal lesions may be difficult to distinguish clinically from melanomas. Many eyes enucleated for suspected melanoma do not have malignant melanomas, but detachments of the uvea, sensory

108   The Eye

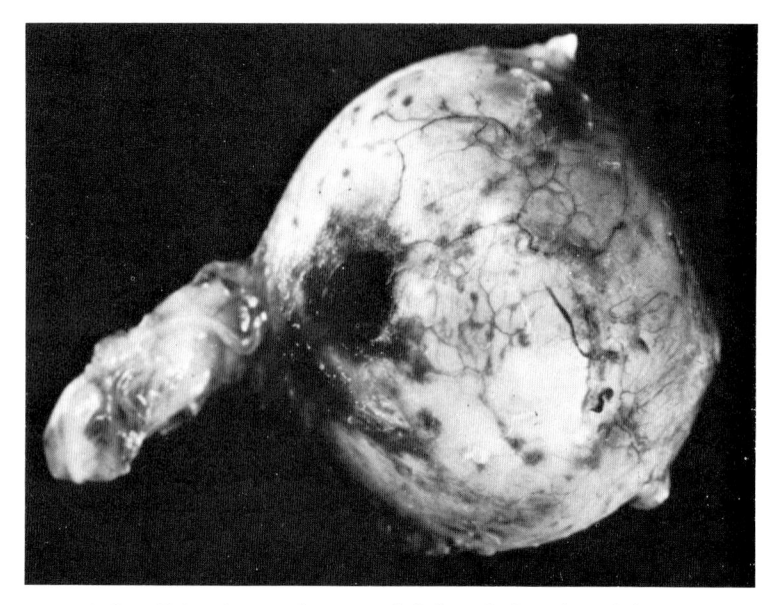

Figure 6.12. A choroidal melanoma has extended through the sclera of this eye. Transcleral extension is a frequent mode of spread with ocular melanomas.

retina, or retinal pigment epithelium, hemangiomas, nevi, retinal pigment epithelial hyperplasia, or metastatic carcinoma.

Melanomas of the Eyelid

Melanomas of the eyelid are rare. They can arise from melanocytes in the normal epidermis or within pre-existing intradermal or junctional pigmented nevi. Melanomas of the eyelid may be divided into superficial spreading, nodular and lentigo-maligna varieties. The superficial spreading malignant melanoma is the most common type, whereas the nodular melanoma has the worst prognosis. Lentigo-maligna melanoma (Hutchinson's melanotic freckle) is a relatively uncommon form of melanoma that most often appears in elderly individuals. The lesion has an irregular outline and surface topography and a characteristic variegated pigmentation. If often becomes much larger than the superficial spreading type. This type of melanoma tends to grow slowly and may remain superficial for 30 years or longer. It yields relatively few metastases. The condition is probably not fatal in about 80–85% of affected individuals.

"Acquired Melanosis" of the Conjunctiva

Occasionally, at about the age of 40 to 50, irregular areas of pigmentation appear spontaneously in a nonpigmented portion of the conjunctiva of one eye. The pigmentation may involve the conjunctiva diffusely or affect multiple discrete areas. The lesions have a variable morphology which ranges from a junctional nevus to an intraepithelial melanoma with variable degrees of invasion. A variety of terms have been applied to this condition which is analogous to lentigo-maligna melanomas of the skin. The condition may regress spontaneously or after a variable period of time (many months to 10 years or longer) it may evolve into a malignant melanoma. The behavior of these lesions is difficult if not impossible to predict, and after local excision, recurrences are common.

Malignant Melanoma of Conjunctiva

An invasive malignant melanoma of the conjunctiva may be preceded by an intraepithelial melanoma, a junctional nevus, or no overt antecedent lesion. Melanomas arising from junctional nevi usually occur at the limbus or inner canthus. Some conjunctival melanomas represent an extension of an intraocular melanoma.

METASTATIC NEOPLASMS

Metastatic neoplasms reach the eye, eyelid, and orbit by way of the bloodstream from many sites, especially from the lung in males and the breast in females. Most occur in the 5th decade, as part of terminal widespread metastases, but sometimes as the presenting clinical manifestation of the cancer. The average interval between the time of detection of the primary and evidence of intraocular metastases is 2 years. The duration of life after enucleation for a metastatic intraocular neoplasm is usually less than 1 year, but may be as long as 9 years.

Neuroblastoma frequently metastasize to the orbit in infancy and childhood.

At one time, it was thought that the left eye was more often involved than the right, but this is not true. Most intraocular metastases involve the highly vascular choroid and especially the posterior portion. Choroidal metastatic neoplasms are typically flat, but may cause a shallow retinal detachment. Unlike melanomas, they seldom penetrate Bruch's membrane. Metastatic carcinoma of the retina is rare. Metastatic carcinoma is much less frequently recognized in the orbit than in the eye.

Cutaneous melanomas rarely metastasize to the eye. When they do, multiple flat lesions usually develop in both eyes and multiple tumor emboli are often present.

Malignant neoplasms of the eyelid, conjunctiva, paranasal sinuses, nose, nasopharynx, and intracranial cavity may involve the orbit.

MISCELLANEOUS NON-NEOPLASTIC MASS LESIONS

Within the ocular tissues, fluid loculates in cysts or pseudocysts. Traditionally, a cyst possesses an epithelial or endothelial lining, whereas a pseudocyst lacks a cellular lining. Cysts may enlarge and displace adjacent structures, like a neoplasm and other mass lesions. Some cysts arise by the abnormal dilation of pre-existing tubules, or ducts, as after an obstruction to the excretory ducts of glands by inspissated secretions or scar tissue (retention cysts). Retention cysts of the conjunctiva are common. Small retention cysts usually occur in the upper and lower conjunctival fornices. Minute yellowish to white concretions accumulate in the conjunctival epithelium especially in the elderly and in individuals with chronic conjunctivitis and may cause retention cysts.

Sudoriferous cysts occasionally occur on the eyelid, usually at the lid margin. They are lined by two layers of epithelial cells, surrounded by a delicate connective tissue capsule, and are filled with a clear fluid. Sudoriferous cysts are thought to arise from obstruction to the excretory ducts of the modified sweat glands of Moll. They may reach several millimeters in diameter and compress adjacent structures.

Single or multiple small intraepithelial cysts commonly occur in the skin of the eyelid (*milia*). They are about 1 mm in diameter and appear as white, round, slightly elevated lesions. They contain keratin and sebaceous material. Most are derived from obstructed hair follicles.

The term *sebaceous cyst* is loosely applied to cutaneous cysts derived from the entire pilosebaceous apparatus, rather than from the sebaceous gland alone. These pilar cysts are lined by keratinized stratified squamous epithelium and rarely are residual sebaceous glands evident in a segment of the cyst lining. Such cysts are common in regions where there are large and numerous hair follicles such as the eyebrows. They may appear as white, solid, or yellow elevations of the skin. They are usually filled with sebaceous material and keratin. Minute pearly white cysts derived from sebaceous glands in the tarsal plate (Meibomian glands) or associated with the cilia (Zeis' glands) are rare. They are usually secondary to obstruction in the lids and are lined by a single or double layer of flattened epithelium. Like other pilar cysts, they contain sebum and keratin.

A dilation of the lacrimal sac (*mucocele of the lacrimal sac*, *hydrops of the lacrimal sac*) results from an obstruction to its drainage. The lacrimal canaliculus may also dilate. A cystic dilation of the canaliculi or of the lacrimal sac may extend into the adjacent orbit and lid.

Mucoceles of the paranasal sinuses are usually the result of an inflammatory obstruction to the ostia through which the sinuses drain

Figure 6.13. These white cysts of the ciliary body contain abundant protein which coagulated when the eye was fixed in formalin. In the unfixed eye, the cysts contained crystal clear fluid.

into the nose. A mucocele of the frontal or ethmoidal sinus may encroach upon the orbit resulting in proptosis.

Conjunctival implantation cysts lined by a conjunctival epithelium may follow ocular trauma or surgery. Epidermoid implantation cysts lined by stratified squamous epithelium, which may be keratinizing, occur in the orbit or eyelids. In contrast to dermoid cysts, their lining lacks skin appendages. They contain keratin and desquamated cells, but lack sebaceous material. Orbital inclusion cysts sometimes follow enucleation ("socket cysts").

In adults, a separation of the two epithelial layers of the ciliary epithelium commonly results in cysts (*cysts of the pars plana*) which have a predilection for the zone adjacent to the ora serrata, especially in the temporal aspect of the globe. Their incidence increases with age. Cysts of the ciliary body can extend into the posterior chamber and are occasionally lightly pigmented. The cysts usually contain crystal clear fluid and acid mucopolysaccharide which is believed to be hyaluronic acid. In individuals with multiple myeloma and other dysproteinemias these cysts contain macroglobulins and other proteins which precipitate after formalin fixation giving the cysts a white opaque appearance (Fig. 6.13).

REFERENCES

ALLEN, R. S., AND STRAATSMA, B. R. Ocular involvement in leukemia and allied disorders. Arch. Ophthalmol. 66:490–508, 1961.
ASHTON, N. Primary tumours of the iris. Br. J. Ophthalmol. 48:650–668, 1964.

ASHTON, N., AND MORGAN, G. Embryonal sarcoma and embryonal rhabdomyosarcoma of the orbit. J. Clin. Pathol. 18:699–714, 1965.

BARR, C. C., GREEN, W. R., PAYNE, J. W., KNOX, D. L., JENSEN, A. D., AND THOMPSON, R. L.: Intraocular reticulum-cell sarcoma: clinicopathologic study of four cases and review of the literature. Survey Ophthalmol. 19:224–239, 1975.

BERARD, C. W. Lymphoreticular disorders-malignant proliferative response-lymphoma. *In* Hematology. Ed. by W. J. Williams, E. Beutler, A. J. Ersley, and R. W. Rundles. Mc-Graw-Hill, Inc., New York, 1972.

BONIUK, M. (Ed). Ocular and Adnexal Tumors. C. V. Mosby Co., St. Louis, 1964.

BONIUK, M., AND ZIMMERMAN, L. E. Eyelid tumors with reference to lesions confused with squamous cell carcinoma. III. Keratoacanthoma. Arch. Ophthalmol. 77:29–40, 1967.

BONIUK, M., and ZIMMERMAN, L. E. Sebaceous carcinoma of the eyelid, eyebrow, caruncle, and orbit. Trans. Am. Acad. Ophthalmol. Otolaryngol. 72:619–642, 1968.

BURKITT, D., AND O'CONNOR, G. T. Malignant lymphoma in African children: I. A clinical syndrome. Cancer 14:258–269, 1961.

DAVIS, F. A. Primary tumors of the optic nerve (a phenomenon of Recklinghausen's disease). A clinical and pathologic study with a report of five cases and a review of the literature. Arch. Ophthalmol. 23:735–821, 957–1022, 1940.

ELLSWORTH, R. Tumors of the Retina. *In* Retinal Diseases in Children, Chap. 3, pp. 39–58. Ed. by W. Tasman. Harper & Row, Publishers, New York, 1971.

FERRY, A. P. and FONT, R. L. Carcinoma metastatic to eye and orbit: I. Clinicopathologic study of 227 cases. Arch. Ophthalmol. 92:276–286, 1974.

FONT, R. L., NAUMANN, G., AND ZIMMERMAN, L. E. Primary malignant melanoma of the skin metastatic to the eye and orbit: Report of ten cases and review of the literature. Am. J. Ophthalmol. 63:738–754, 1967.

FONT, R. L., SPAULDING, A. G. AND ZIMMERMAN, L. E. Diffuse malignant melanomas of the uveal tract: a clinicopathological report of 54 cases. Trans. Am. Acad. Ophthalmol. Otolaryngol. 72:877–895, 1968.

FONT, R. L., ZIMMERMAN, L. E., AND ARMALY, M. F. The nature of the orange pigment over a choroidal melanoma—histochemical and electron microscopic observations. Arch. Ophthalmol. 91:359–362, 1974.

FORREST, A. W. Tumors following radiation about the eye. Trans. Am. Acad. Ophthalmol. Otolaryngol. 65:694–717, 1961.

FORREST, A. W. Pathologic criteria for effective management of epithelial lacrimal gland tumors. Am. J. Ophthalmol. 71:178–192, 1971.

GARNER, A. Tumours of the retinal pigment epithelium. Br. J. Ophthalmol. 54:715–723, 1970.

GASS, J. D. M. Differential Diagnosis of Intraocular Tumors. C. V. Mosby, St. Louis, 1974.

GODTFREDSEN, E. Ophthalmo-neurological symptoms in connection with nasopharyngeal tumours. Br. J. Ophthalmol. 31:78–100, 1947.

HENDERSON, J. W. Orbital Tumors. W. B. Saunders Co., Philadelphia, 1973.

JOHNSON, B. L. Proteinaceous cysts of the ciliary body. I. Their clear nature and immunoelectrophoretic analysis in a case of multiple myeloma. Arch. Ophthalmol. 84:166–170, 1970.

JOHNSON, B. L. Proteinaceous cysts of the ciliary epithelium. II. Their occurrence in nonmyelomatous hypergammaglobulinemic conditions. Arch. Ophthalmol. 84:171–175, 1970.

KWITKO, M. L., BONIUK, M., AND ZIMMERMAN, L. E. Eyelid tumors with reference to lesions confused with squamous cell carcinoma. I. Incidence and errors in diagnosis. Arch. Ophthalmol. 69:693–697, 1963.

NAUMANN, G., YANOFF, M., AND ZIMMERMAN, L. E. Histogenesis of malignant melanomas of the uvea. I. Histopathologic characteristics of nevi of the choroid and ciliary body. Arch. Ophthalmol. 76:784–796, 1966.

O'GRADY, R. B., ROTHSTEIN, T. B., and ROMANO, P. E. D-Group depletion syndromes and retinoblastoma. Am. J. Ophthalmol. 77:40–45, 1974.

PORTERFIELD, J. F., AND ZIMMERMAN, L. E. Rhabdomyosarcoma of the orbit: a clinicopathologic study of 55 cases. Virchows Arch. Pathol. Anat. 335:329–344, 1962.

REESE, A. B. Tumors of the Eye, 2nd edition. New York Hoeber Medical Division, Harper & Row, Publishers, New York, 1963.

SHIELDS, J., AND ZIMMERMAN, L. E. Lesions simulating malignant melanoma of the posterior uvea. Arch. Ophthalmol. 89:466–471, 1973.

Tso, M. O. M., ZIMMERMAN, L. E., AND FINE, B. S. The nature of retinoblastoma. I. Photoreceptor differentiation: A clinical and histopathologic study. Am. J. Ophthalmol. 69:339–349, 1970.

YANOFF, M., and ZIMMERMAN, L. E. Histogenesis of malignant melanomas of the uvea. II. Relationship of uveal nevi to malignant melanomas. Cancer 20:493–507, 1967.

ZIMMERMAN, L. E. (Ed). Tumors of the Eye and Adnexa, Vol. 2, No. 2. International Ophthalmology Clinics, Little, Brown & Co., Boston, 1962.

ZIMMERMAN, L. E. Clinical pathology of iris tumors. Amer. J. Clin. Pathol. 39:214–228, 1963; [Am. J. Ophthalmol. 56:183–195, 1963]

ZIMMERMAN, L. E. Melanocytes, melanocytic nevi, and melanomas. Invest. Ophthalmol. 4:11–41, 1965.

7

DISEASES DUE TO MICROORGANISMS

A wide variety of defense mechanisms protect the eye against the microorganisms with which it comes in contact. These include the barrier provided by the corneal and conjunctival epithelia, polymorphonuclear leukocytes, macrophages, and interferon. Tears wash away potentially pathogenic microorganisms. They also contain the enzyme lysozyme, which possesses bacteriolytic activity against certain bacteria of low pathogenicity, and relatively more immunoglobulin A than the serum. In addition, the lymphoid system combats infections by the production of antibodies and immune cells. Depending on the nature of the infecting agent, either humoral or cellular immunity, or both, may participate in the destruction of infected cells.

Like infections elsewhere in the body, disorders of the immune response predispose to ocular infections. Patients with lymphomas exhibit a low antibody titer and an impaired cell-mediated immunity and are notoriously prone to tuberculosis, as well as certain fungal, viral, and protozoal infections. Corticosteroids and other immunosuppressive drugs result in an increased susceptibility to infection. Stasis, as produced by concretions or stenosis of the nasolacrimal duct or lacrimal sac, may predispose to dacryocystitis and conjunctivitis. Hypogammaglobulinemia, diabetes mellitus, agranulocytosis, and malnutrition predispose to infection. A lesion produced by one pathogenic agent such as the trachoma agent may predispose to infection by others, like bacteria.

Numerous microorganisms produce ocular disease and may reach the eye by several routes. Some, such as *Treponema pallidum*, *Toxoplasma gondii*, and cytomegalovirus, can cross the placenta and infect the unborn child. During the passage through the birth canal, the newborn infant's eyes may become infected by organisms such as herpesvirus (type II), *Chlamydia oculogenitalis*, and *Neisseria gonorrhoeae*. Airborne organisms from dust or sputum may enter the lacrimal system, conjunctiva, or eyelids. Microorganisms can be introduced into the eye by fingers, contaminated eye drops, foreign bodies, or water, as in swimming pools. Bacterial and fungal infections may complicate surgical procedures to the eye, such as cataract extractions and filtering operations for glaucoma, and then wound healing is delayed.

On occasion an infected focus may disseminate to the eye by the bloodstream. Infected emboli usually come from the lungs or heart valves. A neighboring infection in the skin or paranasal sinuses may reach the eye or orbit by direct spread.

The portal of entry for some microorganisms that cause systemic infections like adenoviruses, rubeola, *T. pallidum*, and *Mycobacterium tuberculosis*, may be the conjunctival mucosa. Spread of organisms by the orbital veins into the cavernous sinus can cause a cavernous sinus thrombosis. Lymphatic dissemination may occur from infections of the conjunctiva, upper lid, and outer one-third of the lower lid to the preauricular lymph nodes, and from the medial two-thirds of the lower lid to the submaxillary lymph nodes. Tuberculosis, tularemia, coccidioidomycosis, syphilis, lymphogranuloma venereum, cat scratch fever, and chancroid can result in a chronic, unilateral, necrotic lesion of the conjunctiva or eyelid and a preauricular lymphadenitis which may be ulcerative (Parinaud's oculoglandular syndrome).

BACTERIAL DISEASES

Gram-positive Cocci

Staphylococcus aureus is particularly pathogenic to neonatal and elderly individuals. The hallmark of staphylococcal infections is tissue destruction with abscess and scar formation. The lesions include boils, carbuncles, and impetigo. The eyelids are frequently affected. A localized suppurative infection of the sebaceous glands of Zeis and the associated hair follicles into which they open is common. An acute infection involving a Meibomian gland of the tarsal plate is much more painful than the ordinary sty because it is encased in fibrous tissue deep in the lid. It usually discharges through the conjunctival surface of the lid, but sometimes drains externally. Staphylococcal blepharitis is often associated with chronic inflammation of the conjunctiva and cornea. Minute marginal ulcers may ensue, mainly in the inferior half of the cornea. Some manifestations of staphylococcal blepharitis, such as phlyctenular keratoconjunctivitis, may represent delayed hypersensitivity to staphylococcal antigens.

Erysipelas, a self-limited form of streptococcal skin infection, can involve the eyelid. It is characterized by large, sharply demarcated areas of cellulitis. Marked hyperemia, edema, and occasionally vesicles and hemorrhage occur. The β hemolytic streptococci produce pseudomembranous conjunctivitis with a tendency to involve the cornea. Group A streptococci and *Streptococcus viridans* can cause bacterial endocarditis, which may embolize to various tissues including the eye.

Pneumococcus usually infects the cornea and conjunctiva after trauma and produces ulcers that have a dirty gray color with overhanging margins (serpiginous ulcer). The ulcerated cornea becomes markedly thin and it commonly perforates. *Pneumococcus* may cause dacryocystitis.

Gram-negative Cocci

N. gonorrhoeae is still an important cause of blindness in some parts of the world. The organisms cause a severe, acute purulent conjunctivitis which is commonly accompanied by corneal ulceration and perforation. If untreated, blindness commonly follows panophthalmitis or corneal scarring. The infant usually becomes infected during the passage through the birth canal of an infected mother. Hematogenous dissemination of *N. gonorrhoeae* is rare even in untreated cases of gonorrhea, but uveitis occasionally complicates genital lesions.

Mycobacteria

Mycobacterium tuberculosis

Primary tuberculosis occurs most frequently in children and the initial site of infection is occasionally the conjunctiva. When this occurs, the preauricular lymph nodes become involved. Minute tubercles may develop in the choroid and be the first clinical indication of miliary tuberculosis. There is a low incidence of ocular tuberculosis in patients with active pulmonary tuberculosis. The retina is rarely affected directly, but may be edematous over a uveal granuloma. The skin of the eyelid can be involved in lupus vulgaris and rarely in other forms of cutaneous tuberculosis. Many examples of phlyctenular keratoconjunctivitis seem to be a delayed hypersensitivity to antigens of *M. tuberculosis*. Tuberculous meningitis may result in secondary optic atrophy.

Mycobacterium fortuitum

Acid-fast bacilli have been identified in several chronic corneal ulcers from which *M. fortuitum* was isolated. Most of these lesions have followed physical injuries to the cornea.

Mycobacterium leprae

In certain parts of the world, including Central Africa and India, there is a high incidence of leprosy. Ocular lesions are common in advanced lepromatous leprosy, and almost all untreated cases eventually develop ocular involvement if the disease is of sufficient duration. In leprosy, the eyes are usually bilaterally and symmetrically involved, and there is a predilection for the conjunctiva, cornea, and anterior sclera.

In tuberculoid leprosy, the trigeminal and facial nerves are frequently affected. Involvement of the trigeminal nerve produces anesthesia of the cornea and conjunctiva and eventually causes a neuropathic keratopathy. The facial nerve lesions result in atrophy of the orbicularis oculi, leading to failure of the lids to close and cover the eye, and exposure keratopathy.

Gram-negative Bacilli

Hemophilus influenzae causes an acute mucopurulent conjunctivitis, especially in hot climates. The infection is common particularly in young children. Epidemics may result from eye to eye transmission. *Hemophilus ducreyi* is the causative agent of the chancroid (soft chancre). Occasionally, the conjunctiva and/or eyelid is involved.

Pseudomonas aeruginosa, sometimes introduced into the eye by contaminated fluorescein solutions, causes an acute, rapidly progressive corneal ulcer, which usually begins centrally and may perforate within 48 hours. *Klebsiella pneumoniae* may localize in the eye and cause a suppurative process like that caused by other pyogenic organisms.

Brucellaceae

Tularemia is a disease of rodents caused by *Francisella tularensis* (*Pasteurella tularensis*). It can be transmitted to man by the tick and other vectors from infected small animals. The conjunctiva or eyelid may be the primary site of inoculation (oculoglandular tularemia). When this occurs, a nodule develops which suppurates and ulcerates after an incubation period of 1-10 days. It may be accompanied by chemosis and marked preauricular lymphadenopathy. Another species of gram-negative coccobacilli of the family Brucellaceae causes brucellosis (Malta fever). In brucellosis, uveitis is common and vitreal and retinal hemorrhages also have been described.

Shigella

An iridocyclitis, often associated with polyarthritis, is common 2-4 weeks after bacillary dysentery. Acute and chronic chorioretinitis have been described in chronic bacillary dysentery.

Corynebacteria

A pseudomembranous conjunctivitis may be the primary lesion in diphtheria. The nerves which supply the extrinsic muscles of the eye and the muscles of accommodation are commonly involved.

Gram-positive Bacilli

Clostridia

The pathogenic clostridia, *Clostridium perfringens*, *Clostridium tetani*, and *Clostridium botulinum* rarely affect the ocular tissues. Almost all known cases of clostridial panophthalmitis have followed perforating ocular injuries. *C. perfringens* produces a severe necrotizing reaction associated with the production of gas.

VIRAL DISEASES

A wide variety of viruses affect the ocular tissue (Table 7.1).

Herpesviruses

Herpes Simplex

The herpes simplex virus exists in at least two antigenic types. Primary infection by herpes simplex usually occurs in childhood between the age of 6 months and 5 years. The conjunctival and corneal epithelium are common primary foci of infection by herpes simplex virus type I. Herpes simplex type II affects mostly the genitalia and the skin below the waist. It rarely causes ocular infection, except in the neonate who becomes infected during passage through the birth canal of a mother harboring genital herpes. Type II herpes can produce widespread infection of the cornea and retina. The primary lesions tend to be localized but are often accompanied by a regional lymphadenopathy, systemic infection, and fever. Most lesions are asymptomatic plaques of diseased epithelial cells containing replicating virus, usually healing without ulceration. Less than 10% of primary herpes keratoconjunctivitides progress to corneal ulceration or stromal lesions. The lesions shed the herpesvirus and may infect another individual. An acute unilateral follicular conjunctivitis may occur. Most cases of primary infection are subclinical or undiagnosed. After the host recovers, the virus remains latent.

With secondary herpes infection (reactivation disease), there is a high rate of recurrences at one site. The herpes simplex virus causes numerous

TABLE 7.1
Viruses Affecting Ocular Tissues

Virus	Disease	Ocular lesions
Deoxyribonucleic acid viruses		
Papovaviruses		
Human wart virus	Verruca vulgaris	Warts on eyelids, verrucose conjunctivitis, verrucose keratitis
	Verruca plana	
Adenoviruses		
Type 3		Pharyngoconjunctival fever
Type 8		Epidemic keratoconjunctivitis
Types 1–11, 14–17, 20, 22, 26, 27		Follicular conjunctivitis, keratoconjunctivitis
Herpesviruses		
Herpes simplex	Chickenpox, Herpes zoster	Keratitis, uveitis
Herpesvirus varicellae (varicella-zoster)		Keratoconjunctivitis, scleritis, iridocyclitis, and optic neuritis
Cytomegalovirus	Cytomegalic inclusion disease	Chorioretinitis, dacryocystitis
Epstein-Barr virus	Burkitt's lymphoma	Orbital lymphoma
Poxviruses		
Variola	Smallpox	Catarrhal conjunctivitis, keratitis
Vaccinia		Blepharoconjunctivitis, purulent conjunctivitis, keratitis
Molluscum contagiosum		Molluscum contagiosum, keratoconjunctivitis
Ribonucleic acid viruses		
Picornaviruses		
Rhinovirus of cattle	Foot and mouth disease of cattle	Severe keratoconjunctivitis
Togaviruses		
Rubella virus[a]	German measles	Catarrhal or follicular conjunctivitis, keratitis, cataracts
Arboviruses		
Yellow fever virus	Yellow fever	Intense conjunctival injection, subconjunctival hemorrhages
Dengue fever virus	Dengue fever	Conjunctivitis
Sandfly fever virus	Sandfly fever	Conjunctivitis
Myxoviruses		
Influenza viruses	Influenza	Acute catarrhal or follicular conjunctivitis
Paramyxoviruses		
Rubeola virus	Red measles	Mucopurulent conjunctivitis, Koplik's spots, punctate keratitis
	Subacute sclerosing panencephalitis	Maculopathy, chorioretinitis, optic atrophy, papilledema
Mumps virus	Epidemic parotiditis	Conjunctivitis, episcleritis, keratitis
Newcastle disease virus		Conjunctivitis
Enteroviruses		
Coxsackie		Conjunctivitis

[a] The taxonomic assignment of the rubella virus has not yet been finalized.

minute, discrete intraepithelial ulcers (superficial punctate keratopathy). Some of these lesions heal, others enlarge and eventually coalesce to form linear or branching fissures (dendritic ulcers) (from Greek *dendron*, a tree) (Fig. 7.1). The epithelium between the fissures desquamates, causing sharply demarcated, irregular geographic ulcers. The ulceration may be demonstrated in the patient by staining the cornea with fluorescein. The affected epithelial cells contain eosinophilic intranuclear inclusion bodies (Lipschütz bodies) and may become multinucleated. Such cells may be present in corneal epithelial scrapings. The ulceration appears after the serum antibodies are elevated.

Recurrences may be precipitated by ultraviolet light, trauma, menstruation, emotional and physical stress, exposure to light or sunlight, vaccination, and other factors. Why these recurrences occur remains to be explained. Because the virus can reside in the trigeminal ganglion and possibly in the lacrimal gland, recurrent infection may follow extension down the nerves or extrusion in the tears. These infections occur despite high titers of circulating, highly stable, neutralizing antibody and cell-mediated immunity. Unlike primary infection, ulceration and a more severe inflammatory reaction occur, whereas fever and lymphadenopathy are not features. A leukocytic and vascular invasion of the cornea may occur.

Different types of lesions occur in the corneal stroma. A central disc-shaped corneal opacity sometimes develops beneath the epithelium

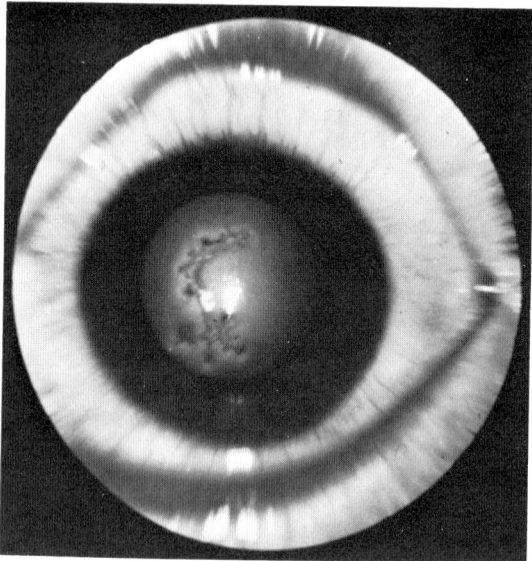

Figure 7.1. Dendritic ulcer of the central cornea.

several days after herpes keratitis (disciform keratitis) due to edema associated with a minimal inflammatory cell infiltrate. The virus can invade the stroma and occasionally the eye. After chronic postherpetic stromal alterations, Descemet's membrane may bulge into the cornea (descemetocele) and corneal perforation can occur.

Cytomegalovirus

Cytomegalovirus causes lesions characterized by greatly enlarged cells containing intranuclear and intracytoplasmic inclusion bodies. A necrotizing chorioretinitis may develop in infants and be accompanied by microcephaly and periventricular calcification. Chorioretinitis due to this virus also occurs in children and adults with malignant disorders of the hematopoietic system, in individuals with debilitating disease, and also in patients with an abnormality of the immune mechanism, either a primary defect or one secondary to long term use of immunosuppressive drugs.

Herpesvirus Varicellae

Herpesvirus varicellae produces two clinically dissimilar diseases. Chickenpox (varicella) is believed to be the response in the nonimmune host, whereas herpes zoster occurs in a partially immune host. Chickenpox, an acute, contagious disease, may involve the eyelids, but rarely the conjunctiva and cornea. Herpes zoster, uncommon in children, occurs most often in patients over the age of 50. Several conditions predispose to herpes zoster: trauma, drug therapy, tuberculosis, and malignancy (especially lymphomas). It is believed that the virus spreads to the skin from the dorsal root ganglia by way of the nerves. Herpes zoster is thought to be due to reactivation of dormant herpesvirus varicellae and seldom due to reinfection. In herpes zoster, the ganglia of the sensory nerves are involved. A cutaneous vesicular eruption, similar to that of varicella, develops in dermatomes innervated by affected sensory nerves. The ophthalmic division of the trigeminal nerve is the commonest cranial nerve affected (Fig. 7.2) and involves the eyelid. Lesions on the tip of the nose accompany involvement of the nasociliary branch of the ophthalmic nerve and often precede lesions of the cornea. Diminished corneal sensation, accompanied by edema of the cornea, commonly develops. A necrotizing inflammation of the uvea and hyphema may occur.

Papilloma and Papovaviruses

The common warts, verruca vulgaris and verruca plana, frequently occur on the eyelids. They are transmitted by autoinoculation and by

122   The Eye

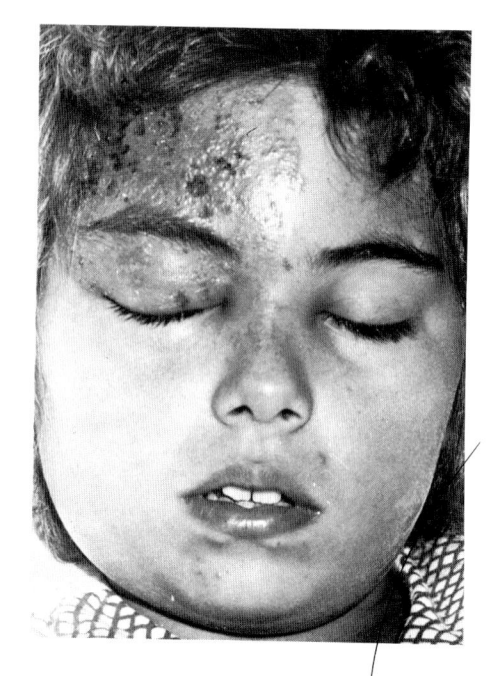

Figure 7.2. Herpes zoster. A vesicular eruption of the skin involves the ophthalmic division of the trigeminal nerve.

contact and are believed to be due to identical or closely related deoxyribonucleic acid viruses.

Adenoviruses

Many strains of adenovirus cause an acute follicular conjunctivitis. Conjunctival hyperemia, chemosis, and a preauricular lymphadenitis often accompany the infection. Conjunctival membranes and pseudomembranes are commonly due to adenovirus types 6, 7, and 8. With adenovirus type 8, the cornea is conspicuously involved (epidemic keratoconjunctivitis). Central or paracentral nummular opacities appear in the subepithelial corneal tissue 7-10 days after the onset of a transient conjunctival pseudomembrane and preauricular lymphadenopathy. The keratoconjunctivitis usually resolves completely. In children, a follicular conjunctivitis is accompanied by upper respiratory, gastrointestinal, and other systemic manifestations. In individuals of all ages, but especially children, some adenoviruses (especially type 3) cause a mild nasopharyngitis and conjunctivitis, often with regional lymphadenopathy which lasts 1-2 weeks (pharyngoconjunctival fever). The condition occurs sporadically or as an epidemic, especially in the summer, when it is believed to be transmitted in swimming pools. The conjunctivitis associated with the follicular hyperplasia (especially in the lower cul-de-sac) may be monocular and purulent.

Rubella Virus

A conjunctivitis which is usually bilateral is common in rubella (German measles). Unlike rubeola, rubella spares the tarsal area and the cornea. Ocular and systemic abnormalities (including cataracts, prematurity, failure to thrive, stunted growth, cardiovascular malformations, microcephaly, and deafness) occur in infants born of mothers who develop rubella during the first trimester of pregnancy. A dense, nuclear cataract often develops in a swollen spherical lens. A distinctive but nonspecific feature of the cataract is the retention of karyorrhetic or pyknotic nucleated lens fibers within the lens nucleus. The cortical changes are variable. Chronic iridocyclitis, often with necrosis of the ciliary epithelium, may occur. Retinal pigment epithelial changes, with alternating areas of hypo- and hyperpigmentation, occur. Vacuolization of the pigment epithelium of the iris, hypoplasia of the iris, congenital glaucoma, and microphthalmos with a cloudy cornea may be evident. Degeneration and necrosis of the pigment epithelium of the ciliary body is associated with pigment-laden macrophages in the underlying tissue. A pigmentary retinopathy is often most marked at the posterior pole. The rubella virus may persist in the infant for many months and has been isolated from cataractous lenses even 3 years after birth.

Molluscum Contagiosum

Molluscum contagiosum may infect the epidermis of the eyelid and lid margins, as well as the skin in other areas. Single or multiple yellowish pink, hyperplastic, umbilicated papules, 1–10 mm in diameter, develop. Affected cells contain eosinophilic inclusions (molluscum or Henderson-Patterson bodies) that are 20–30 μ in diameter (Fig. 7.3). Each inclusion consists of numerous elementary bodies (Lipschütz granules). The condition is spread by direct contact and occurs at any age, but especially in children. Molluscum contagiosum of the eyelids may be associated with a superficial punctate keratopathy, follicular or papillary conjunctivitis, and superficial pannus, as well as scarring which resembles trachoma. The virus has not been identified in the cornea or conjunctiva, and because the lesions in these sites resolve if the lid lesions are excised, it is presumed that they are secondary to an immunologic reaction to extruded viral antigens.

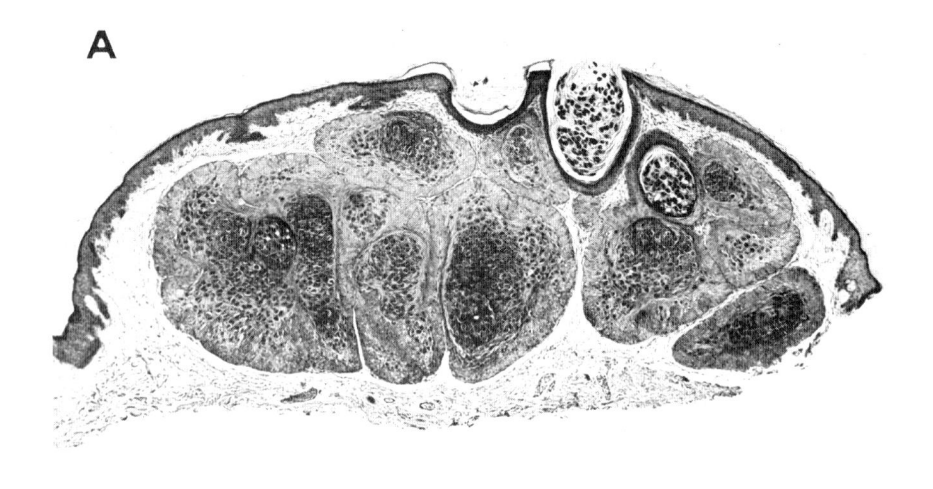

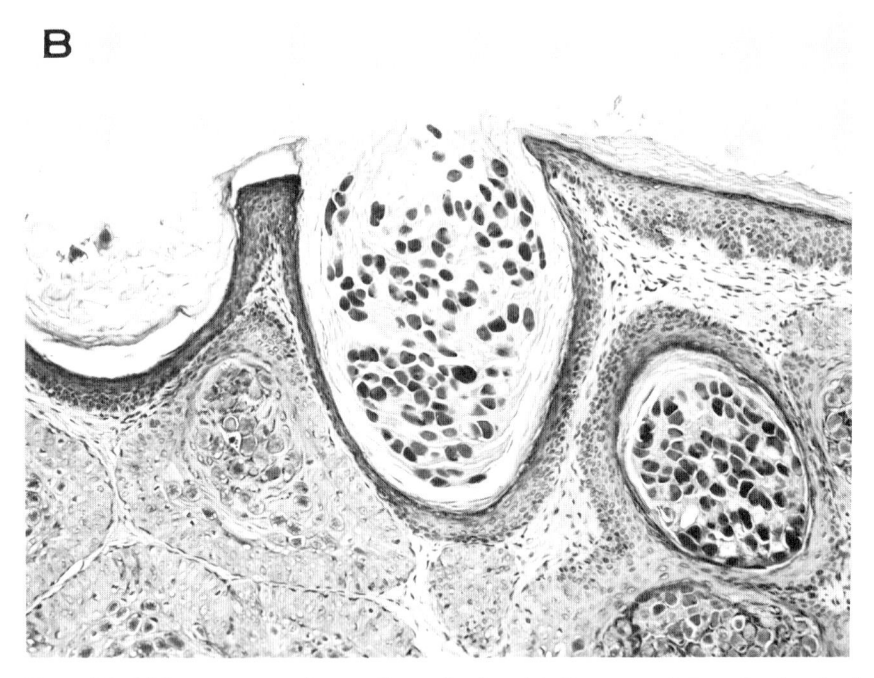

Figure 7.3. Molluscum contagiosum. Hyperplastic epithelium extends into the dermis of this umbilicated papule (*A*). (H & E, × 30) Numerous cytoplasmic inclusions are evident under higher magnification (*B*). (H & E, × 100)

Newcastle Disease Virus

Newcastle disease virus of fowl, an important infection of the poultry industry, can cause conjunctivitis and preauricular lymphadenitis in individuals exposed to infected animals or the virus in the laboratory.

Rubeola (Measles) Virus

A catarrhal conjunctivitis often occurs during the acute phase of measles, and sometimes a nonpurulent keratoconjunctivitis with multiple punctate corneal epithelial erosions develops. Koplik spots involve the mucous membranes, including the conjunctiva, especially the semilunar fold or caruncle. Secondary infection of the eye may result in perforating corneal ulcers and panophthalmitis. In subacute sclerosing panencephalitis, an uncommon manifestation of the measles virus, focal retinal lesions frequently occur, paricularly in the macular area. Although usually discrete, the lesions may be multiple. Blindness resulting from subacute sclerosing panencephalitis commonly appears after the neurologic manifestations, but occasionally can be the first sign of the disease.

Mumps Virus

This acute contagious disease has a predilection for the salivary glands, especially the parotid. Ocular manifestations include transient corneal edema, acute dacryoadenitis, and uveitis.

Variola

In certain parts of the world, smallpox is endemic and is a common cause of blindness. The disease is characterized by a generalized vesiculopapular eruption that heals with scarring. The conjunctiva and cornea may be involved and a catarrhal or purulent conjunctivitis is common.

Vaccinia

In individuals, especially children, vaccinated for smallpox, accidental self-inoculation of the eyelids or the cornea may cause an acute inflammatory reaction associated with pustules. The lesions are characterized by epithelial hyperplasia and eosinophilic intracytoplasmic inclusion bodies (Guarnieri bodies). Superficial corneal ulcerations with undermined edges occur.

DISEASES DUE TO SPIROCHETES

Treponematoses

Spirochetal disease is still common, particularly in some parts of the world. In syphilis, due to *Treponema pallidum*, the chancre may occur on the eyelids or conjunctiva. During the secondary stage of syphilis, when serologic tests are positive, there may be an acute catarrhal conjunctivitis with prominent papillae, but without the formation of lymphoid follicles. A transient hyperemia of the iris (roseola), acute iridocyclitis, and retrobulbar neuritis may occur in the secondary stage.

Most patients with syphilitic iridocyclitis have an associated skin rash. Chorioretinitis is rare, but may occur 10 years after the chancre. When the choroid is involved, the retina undergoes secondary atrophy and gliosis. Ischemic retinal lesions are usually associated with syphilitic angiitis. In tertiary syphilis, gummas can develop in tissues like the uvea, conjunctiva, or eyelids. Meningovascular neurosyphilis may result in papilledema or optic atrophy. In tabes dorsalis, optic atrophy is common. A gumma in the brain rarely gives rise to increased intracranial pressure and papilledema. The pupils may react to accommodation, but not to light (Argyll Robertson pupils). In congenital syphilis, an acute iridocyclitis occurs when the infant is about 6 months old and seems to always precede the interstitial keratitis which develops between the ages of 5 and 20 years. A marked pigmentary proliferation is a feature of congenital syphilitic chorioretinitis, leading to a funduscopic appearance of sprinkled salt and pepper.

Yaws, caused by *Treponema pertenue*, occurs in Equatorial Africa, Asia, South America, parts of Australia, and in the South Pacific Islands. The disease, like syphilis, can be divided into primary, secondary, and tertiary stages. During the secondary stage, the face, including the eyelids and eyebrows, is infrequently involved. A catarrhal conjunctivitis may occur. Keratitis and iritis are rare.

RICKETTSIAL DISEASES

After entering the body by the bite of an arthropod, rickettsiae multiply at the site of inoculation. Later, the organisms disseminate widely by the bloodstream, causing focal lesions in various tissues, including the eye. Rickettsiae have a predilection for the endothelium of capillaries and other small blood vessels and cause a localized endothelial swelling and necrosis with thrombosis. In Rocky Mountain spotted fever and during the acute phase of tropical typhus fever (Tsutsugamushi fever), conjunctival hyperemia with or without hemorrhages may occur. In epidemic louse-borne typhus due to *Rickettsia prowazekii*, vitreous hemorrhages, retinal edema, and papilledema can occur. The conjunctiva may be the site of a primary eschar in Old World tick typhus (*Rickettsia rickettsii*) if blood from an infected crushed tick enters the eye.

FUNGAL DISEASES

Fungi reach the ocular tissue by a variety of different routes. Airborne fungi commonly enter the conjunctival cul-de-sac, and can be accidentally inoculated into the eye. They may also reach the eye by direct spread from the skin, paranasal sinuses, nasopharynx, central nervous system and other neighboring structures, or rarely by hematogenous spread from an infected focus elsewhere. Injuries with plants, such as tree branches, thorns, and splinters frequently inoculate pathogenic fungi. Several factors predispose to fungal diseases. Patients with impaired immunologic defenses, whether as a consequence of a primary disease or its treatment, are particularly susceptible to fungal infections. Endocarditis, due to *Candida* or other fungi, occasionally complicates open heart surgery and drug addiction and sometimes results in emboli to the retina. The use of central venous catheters for intravenous feeding predisposes to systemic candidiasis. *Candida albicans* may cause chorioretinitis and commonly enters the vitreous, causing white opacities. Sometimes, panophthalmitis results. Diabetes mellitus predisposes to orbital phycomycosis (mucormycosis) (Fig. 7.4). Aspergillosis may also involve the orbit. The eyelid is sometimes involved in sporotrichosis, North American blastomycosis, the dermatomycoses, and other fungal diseases. Lesions of the uvea may be caused by *Blastomyces dermatitidis*, *Coccidioides immitis*, and other fungi. *C. immitis* may also involve the retina, conjunctiva, and eyelids.

A wide variety of different fungi (including *Fusarium sp.*) have been cultured from and identified in corneal ulcers. The topical application of broad spectrum antibiotics and/or corticosteroids into the conjunctival sac not only increases the relative virulence of recognized fungal pathogens, but allows other commensal species to become pathogenic. Some keratomycoses are characterized by a fluffy, white elevation with shallow peripheral craters surrounded by a sharply demarcated halo ("immune ring").

Rhinosporidium seeberi, endemic in such areas such as India, gives rise to characteristic translucent polyps of the conjunctiva, eyelid, or lacrimal sac. Intraocular cryptococcosis is uncommon, but may occur as an extension from the brain along the optic nerve or by way of the bloodstream from a localized or disseminated infection. Most cases are

128 The Eye

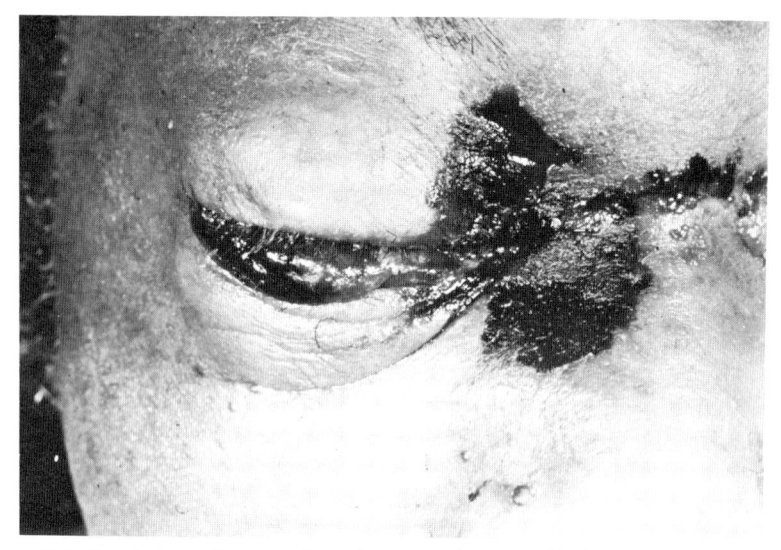

Figure 7.4. Proptosis and a necrotic eschar secondary to orbital mucormycosis in a leukemic patient.

part of systemic cryptococcal disease. *Histoplasma capsulatum* is widely felt to be an important cause of ocular disease in geographic areas where the fungus is endemic, although the organism has only rarely been identified within the eye.

DISEASES DUE TO CHLAMYDIA (BEDSONIAE)

Chlamydia, the obligate intracellular organisms that contain both deoxyribonucleic acid and ribonucleic acid, have been subdivided into two main groups: Group A (trachoma and inclusion conjunctivitis agents) and Group B (which causes psittacosis and other diseases). The trachoma and inclusion conjunctivitis agents manifest their principle pathogenic effect on the eye.

Trachoma

Trachoma is an acute infectious cicatrizing keratoconjunctivitis due to the infectious agent *Chlamydia trachomatis*. It is estimated that several hundred million people are affected by trachoma, the most common cause of blindness in the world. The disease is widespread, but is especially prevalent in regions with a warm, dry climate. It occurs in Asia, the Middle East, and in countries bordering the Mediterranean

Sea. In the United States, the disease is almost restricted to the American Indian in the southwestern part of the country. The pathogen is presumed to be transmitted in several ways: from one eye to the other by fingers, fomites, or flies. Trachoma is not very contagious and a high incidence in a particular population occurs only when factors such as overcrowding and poor hygienic conditions favor its transmission.

Trachoma is the only known cause of follicular conjunctivitis associated with scarring and contraction. The condition is almost always bilateral and usually has an insidious onset. It begins in the conjunctival epithelium of the upper lid and spreads to the deeper tissues including the tarsal plates.

The lesions are characterized by capillary dilation and proliferation in the tarsal conjunctiva. There is a cellular infiltrate which is predominantly lymphocytic, but includes plasma cells and macrophages. Lymph follicles with germinal centers containing histiocytes and lymphoblasts occur. Necrosis of the center of the follicle is characteristic of trachoma. The disease eventually causes an infiltration into the cornea of lymphocytes and blood vessels from the limbus. Initially, this so-called trachomatous pannus, which requires previous sensitization to develop, extends between the corneal epithelium and Bowman's zone. The upper half of the conjunctiva and the cornea are more seriously involved than the lower. Desquamated conjunctival epithelium contains intracytoplasmic inclusion bodies (Halberstaedter-Prowazek). Nuclear fragments occur in large macrophages (Leber cells). Involvement of the Meibomian glands and accessory lacrimal glands results in their atrophy. Scarring of the conjunctiva and tarsal plates leads to distortion of the lids.

In children, where the disease is endemic, it is usually contracted early in life, has a slow course, and spontaneous healing is common. In adults, the disease is more severe, progresses more rapidly, and spontaneous healing is rare. Secondary bacterial infection is common.

Inclusion Conjunctivitis

Chlamydia oculogenitalis, the causative organism of inclusion conjunctivitis (inclusion blennorrhea), can cause conjunctivitis, as well as a low grade urethritis in the male and an asymptomatic endocervicitis in the female. Spread may occur to the eye during passage through the birth canal, by swimming in nonchlorinated pools ("swimming pool conjunctivitis"), or from discharges of ocular, urethral, or cervical lesions. In the neonate, the organism produces an acute purulent conjunctivitis without lymph follicles about 5–10 days after birth (inclusion blennorrhea). In the adult, the organism causes a follicular conjunctivitis (inclusion conjunctivitis) which may persist for 3-12 months if untreated. The intracytoplasmic inclusion bodies are indistinguishable from those in trachoma, but the diseases differ in several aspects. Unlike trachoma, the lower tarsal conjunctival is involved and limbal follicles develop. Scarring does not occur and the follicles do not necrose. Moreover, keratitis is rare and generally mild.

Lymphogranuloma Venereum

The primary lesion in this contagious venereal disease consists of a single, small, painless papule which usually develops into a vesicle that ruptures, giving rise to a shallow gray ulcer with clear-cut edges. It is usually on the external genitalia but may be on the lid, conjunctiva, or elsewhere. It generally heals spontaneously in 7-14 days. The surrounding tissue is discolored, but it is not indurated in contrast to syphilis. The organism spreads by the lymphatics to the regional lymph nodes which are involved from 7-14 days after the primary lesion develops. The lymph nodes frequently suppurate (buboes). Aside from primary lesions of the eyelid or conjunctiva, the ocular tissues may be involved as a result of hematogenous spread.

PARASITIC DISEASES OF THE EYE
Protozoa

Toxoplasma gondii

Based on serologic surveys, toxoplasmosis is contracted by a large percentage of individuals throughout the world. The disease may be acquired by consuming uncooked meat. The coccidian stages of toxoplasma occur in the intestine of certain animals, including the domestic cat, who excrete oocysts in their feces. The oocysts may also infect man and other animals. They may be ingested or inhaled as dust from saliva or dried excreta of infected individuals. After invasion of the host by toxoplasma, the protozoa is carried from the portal of entry to the regional lymph nodes and dispersed throughout the tissues. The organism enters the cells and proliferates in the cytoplasm, generally filling it and displacing the nucleus toward the periphery ("pseudocyst"). During the initial primary infection in a nonimmune host, the virulent parasite destroys infected cells in a few days liberating trophozoites into the tissues. This is followed by their dissemination. After the development of an immunity to the organisms, the extracellular parasites become destroyed, whereas the intracellular merozoites surround themselves by a resilient membrane within which they multiply ("true cysts"). Later, these cysts become extruded from the host cells and lie extracellularly where they may excite little or no reaction. Within these cysts, innumerable toxoplasma may persist for an indefinite period of time. An inflammatory reaction develops around ruptured cysts and necrotized cells in an immune individual. In patients immunosuppressed from an underlying disease or treatment, or both, antibody titers may be extremely low and relapses of toxoplasmosis may occur.

Most primary infections are probably asymptomatic, subclinical illnesses that reflect themselves only by a positive immunologic reaction. The clinical disease may result in a febrile or afebrile lymphadenopathy accompanied by abnormal circulating lymphocytes. When a pregnant woman develops primary infection, she may transmit the organism transplancentally. According to serologic surveys, about one-third to one-half of babies born of women who acquire toxoplasmosis during pregnancy are infected. The infant may appear normal at birth, but later develop overt neonatal disease. In congenital infection, toxoplasma has a predilection for the brain and retina. Ocular involvement is almost always present in neonatal toxoplasmosis and an exudate in the vitreous often obscures the funduscopic appearance of the retinal lesions. This is in contrast to retinal lesions due to cytomegalovirus which are well visualized, because the amount of vitreal exudate is less. Single or multiple foci of chorioretinitis occur predominantly in congenital toxoplasmosis, but may not be evident until 6 months of age. The lesions are usually bilateral. The macula is commonly involved and lesions may be limited to this area (Fig. 7.5). A marked coagulative necrosis of the affected retina occurs in the vicinity of the organisms, which may be within "true cysts," pseudocysts, or be free-lying. The retinal lesions extend into the adjacent choroid, where a granulomatous reaction occurs. Destruction of the affected choroid and retina exposes the white sclera to funduscopy. With marked ocular involvement, microphthalmia may occur.

Mastigophora

The class Mastigophora, or hemoflagellates, contains two genuses that are pathogenic to man: *Leishmania* and *Trypanosoma*. The eyelids may be involved in several varieties of leishmaniasis. In Old World cutaneous leishmaniasis due to *Leishmania tropica*, the lesions are localized to the skin of the uncovered parts of the body and the eyelids are affected in 2–5% of cases. In South American trypanosomiasis caused by *Trypanosoma cruzi*, the primary lesion (chagoma) may develop on the

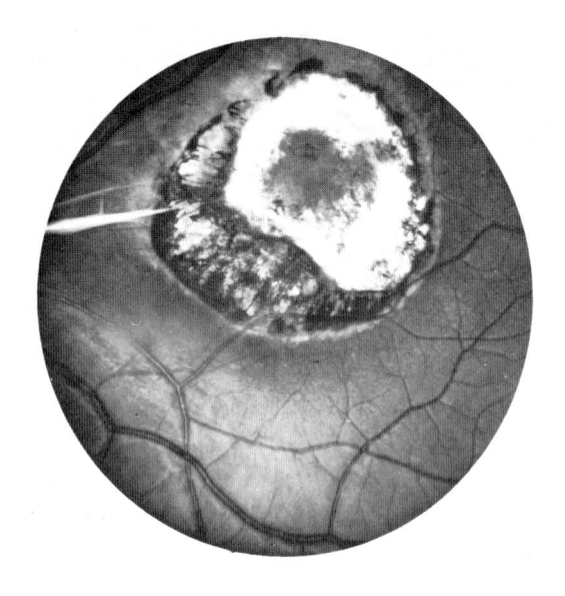

Figure 7.5. Toxoplasmosis of the macula. A large atrophic scar is present with a surrounding rim of pigment epithelial hyperplasia.

eyelids, resulting in a characteristic edema. Urticaria of the eyelids is common in early African trypanosomiasis. When the central nervous system is involved, external ophthalmoplegia, ptosis, papilledema, and optic atrophy may develop.

Ameba

Of the free-living soil amebae two genera are known to cause human disease: *Acanthamoeba* (*Hartmannella*) and *Naegleria*. For some time it has been well established that *Naegleria* sp. causes an acute meningoencephalitis that is usually rapidly fatal. Recently *Acanthamoeba* has become recognized as a cause of progressive indolent corneal ulceration.

Nematodes (Roundworms)

Toxocara

The most frequently identified nematode in the human eye is *Toxocara canis*, a common world-wide intestinal parasite of young dogs. The incidence of toxocariasis varies in different parts of the world. In some areas, it is highly endemic. In a pregnant dog, the larvae of *Toxocara* migrate through the placenta to the upper intestinal tract of the fetus. Here, they develop into mature adult worms which excrete eggs until the worms die when the puppies are about 6 weeks old. The eggs contaminate the hands of children who fondle the infected animal, crawl on the floor, or eat soil contaminated with ova excreted by infected animals. The bitch ingests the ova of toxocara from her offspring during suckling. In the gut of the dog or the human infant, the ova develop into free-swimming larvae. These penetrate through the wall of the intestine into the blood and lymphatics and migrate to the viscera. Larvae in the lungs of the newborn animal migrate via the trachea to the small intestine where they mature to adult worms.

Ocular toxocariasis is a rare cause of unilateral blindness in childhood. It usually presents with visual impairment or strabismus, and may simulate retinoblastoma, as well as other ocular conditions in young infancy. From the clinical standpoint, generalized evidence of visceral larva migrans is rare. There is no apparent sex predilection and most cases have been between 2 and 6 years of age. A chronic endophthalmitis is the most common ocular lesion. A sensory retinal detachment may follow a subretinal exudate or the organization of a vitreal abscess. A solitary granuloma (about 1 disc dimeter in size) may develop in the retina or subretinal space, usually at or near the macula. The peripheral retina is rarely involved. The lesion is associated with a prominent inflammatory cell infiltrate which includes numerous eosinophils. Scarring may be a sequel. Fibrinoid necrosis, with or without a granulomatous reaction, is frequently evident around the larvae. A positive skin sensitivity test, serum antibodies to toxocara, and an eosinophilia of the peripheral blood commonly are present. Because adult worms do not develop in man, the stool examination is unremarkable.

Trichinella

Another nematode parasitic to man is *Trichinella spiralis*. Man becomes infected by eating inadequately cooked meat that contains encysted larvae of this worm. Encysted larvae exist in the muscle of an animal reservoir, usually a pig. Within the intestine, the ingested larvae eventually become sexually mature adults. Eggs develop and hatch in the viviparous female who releases the larvae. Once liberated, the larvae enter the lymphatics and capillaries in the mucosa of the intestine and then pass into the circulation. A swelling of the lids and conjunctiva may occur at this stage. By about the 8th day, the larvae develop only in striated muscle where they coil up and encyst (Fig. 7.6). The extraocular muscles are commonly affected, resulting in tenderness and weakness of ocular movement. The encysted larvae calcify after they have been present in the muscle for about 6 weeks.

ciliated embryos of *Diphyllobothrium* species are ingested by fresh water crustaceans within whose tissues they become small spindle-shaped solid larvae (procercoids). After this host is consumed by a fish or frog, the larvae develop into globular worm-like organisms (plerocercoids) within the tissues of the new host. Human infection is acquired by the ingestion of infected water, fleas, or, as in the Orient, by the application of frog flesh to wounds or mucous membranes, such as the conjunctiva. Sparganum plerocercoids may be present in the conjunctiva, eyelid, and orbit, which may become acutely inflamed after the death of the organism.

Trematodes

A few of these flat, leaf-like globular organisms may affect the ocular tissues. *Schistosoma* rarely causes a granulomatous conjunctivitis after hematogenous dissemination of the ova or due to cercaria which penetrate the eye from infected water. *Paragonimus westermani*, the lung fluke, may rarely deposit the eggs in the orbit and provoke a chronic granulomatous reaction.

Anthropods

Myiasis, an infestation of larvae of dipterous insects which feed on tissues of the host, is rare in North America and Western Europe. It is prevalent where living and sanitary conditions are inadequate and personal hygiene is poor. The most important species are members of the families of Muscidae (houseflies), Calliphoridae (blowflies), Gasterophilidae (horse botflies), and Oestridae (warble flies). The flies deposit eggs or living larvae in neglected superficial ulcers or wounds, as well as on the eyes. The larvae multiply in the superficial tissues of the conjunctiva and eyelid and may penetrate into the lacrimal drainage apparatus or pierce through the conjunctiva and into the eyeball. In at least one species (*Hypoderma bovis*), the eggs may actually be laid in the globe. The larvae may destroy the eyeball and invade the orbit.

An infestation of the eyelid by lice may cause blepharitis or secondary infection of the skin. The hairs of numerous caterpillars are capable of penetrating the surface of the eye and becoming lodged intraocularly (*ophthalmia nodosa*). Caterpillar hairs incite a marked inflammatory reaction to some extent by virtue of being a foreign body, but mainly because of a coating of venom secreted by a gland at the base of the hair.

REFERENCES

ALLEN, H. F., BURNS, R. P., GINGRICH, W. D., GIVNER, S. N., JR., KIMURA, S. J., AND THYGESON, P. Infectious Diseases of the Conjunctiva and Cornea. Symposium of the New Orleans Academy of Ophthalmology. C. V. Mosby, Co., St. Louis, 1963.

ASHTON, N. Toxocara canis and the eye. *In* Corneo-Plastic Surgery, pp. 579-591. Ed. by P. V. Rycroft, Pergamon Press, Oxford, 1969.
BONIUK, M., PHILLIPS, C. A., HINES, M. J., AND FRIEDMAN, J. B. Adenovirus infections of the conjunctiva and cornea. Trans. Am. Acad. Ophthalmol. Otolaryngol. 70:1016-1026, 1966.
BURNS, R. P., AND RHODES, D. H., JR. *Pseudomonas aeruginosa* keratitis: Mixed infections of the eye. Am. J. Ophthalmol. 67:257-262, 1969.
COGAN, D. G., KUWABARA, T., YOUNG, G. F., AND KNOX, D. L. Herpes simplex retinopathy in an infant. Arch. Ophthalmol. 72:641-645, 1964.
DAWSON, C. R. Epidemic Koch-Weeks conjunctivitis and trachoma. Am. J. Opthalmol. 49:801-808, 1960.
DAWSON, C., TOGNI, B., AND MOORE, T. E., JR. Structural changes in chronic herpetic keratitis: Studied by light and electron microscopy. Arch. Ophthalmol. 79:740-747, 1968.
DE VENECIA, G., ZU RHEIN, G. M., PRATT, M. V., AND KISTEN, W. Cytomegalic inclusion retinitis in an adult. Arch. Opthalmol. 86:44-57, 1971.
FEDUKOWICZ, H. B. External Infections of the Eye: Bacterial, Viral, and Mycotic. Appleton-Century-Crofts, New York, 1963.
FENKEL, J. K. Toxoplasmosis: mechanisms of infection, laboratory diagnosis and management *In* Current Topics in Pathology, Vol. 54, pp. 28-75, Springer-Verlag, New York, 1971.
FRANCOIS, J., AND RYSSELAERE, M. *Oculomycoses*, Charles C Thomas, Publisher, Springfield, Ill., 1972.
GRIFFIN, J. R., PETTIT, T. H., FISHMAN, L. S., AND FOOS, R. Y. Blood-borne candida endophthalmitis. Arch. Ophthalmol. 89:450-456, 1973.
HOWARD, G. M., AND KAUFMAN, H. E. Herpes simplex keratitis. Arch. Ophthalmol. 67:373-387, 1962.
KLINTWORTH, G. K., HOLLINGSWORTH, A. S., LUSMAN, P. A., AND BRADFORD, W. D. Granulomatous choroiditis in a case of disseminated histoplasmosis. Arch. Ophthalmol. 90:45-48, 1973.
LANDERS, M. B., III, AND KLINTWORTH, G. K. Subacute sclerosing panencephalitis (SSPE). Arch. Ophthalmol. 86:156-163, 1971.
MATAS, R. B., DAWSON, C. R., AND TOGNI, B. Vaccinia-infected rabbit cornea: a transmission electron microscopic study. Invest. Ophthalmol. 12:782-786, 1973.
NAGINGTON, J., WATSON, P. G., PLAYFAIR, T. J., McGILL, J., JONES, B. R., AND STEELE, A. D. McG. Amoebic infection of the eye. Lancet 2:1537-1540, 1974.
NAUMANN, G., GREEN, W. R., AND ZIMMERMAN, L. E. Mycotic keratitis. Am. J. Ophthalmol. 64:668-682, 1967.
PAVAN-LANGSTON, D., AND McCULLEY, J. P. Herpes zoster dendritic keratitis. Arch. Ophthalmol. 89:25-29, 1973.
PAVAN-LANGSTON, D., AND NESBURN, A. B. The chronology of primary herpes simplex infection of the eye and adnexal glands. Arch. Ophthalmol. 80:258-264, 1968.
ROSS, H. W., AND LAIBSON, P. R. Keratomycosis. Am. J. Ophthalmol. 74:438-441, 1973.
ZIMMERMAN, L. E. Histopathologic basis for ocular manifestations of congenital rubella syndrome. Am. J. Ophthalmol. 65:837-862, 1968.
ZIMMERMAN, L. E. Changing concepts concerning the pathogenesis of infectious diseases. Am. J. Ophthalmol. 69:947-964, 1970.

8

METABOLIC DISORDERS OF THE EYE

Metabolic reactions occur in all diseases, but conditions due to such varied causes as toxins, nutritional deficiencies, mutant genes, and chromosomal abnormalities fundamentally result in a metabolic disorder. Many of the more than 2000 recognized human genetically determined diseases affect the eye (Tables 8.1-8.3). In some of these, the metabolic defect is known, but, in most, it remains obscure.

DISORDERS OF CARBOHYDRATE METABOLISM

Diabetes Mellitus

Diabetes mellitus frequently involves the eye. It is the most common systemic disease that causes blindness. Ocular symptoms occur in 20-40% of diabetics at the clinical onset of the disease and may be a presenting clinical manifestation.

Diabetic Retinopathy

Diabetic retinopathy is the most important lesion that leads to blindness. The retinopathy is characterized by capillary microaneurysms (Figs. 4.5, 4.6), hemorrhages, and exudates. The first discernible, clinical abnormality in diabetic retinopathy may be retinal venous engorgement. The engorged retinal veins may form localized sausage-shaped distensions, coils, and loops. A microangiopathy first appears at the posterior pole, but eventually can involve the entire retina. This stage is followed by small hemorrhages in the same area. Most hemorrhages are in the inner nuclear and outer plexiform layers ("dot and blot hemorrhages"). Large hemorrhages are uncommon. Capillary microaneurysms are a ready source of hemorrhage and exudation. Eventually, glistening, yellow, waxy, circumscribed exudates occur chiefly in the vicinity of microaneurysms and become more numerous with time. Their high lipid content reflects the hyperlipemia that is common in diabetes. In contrast to the juvenile diabetic, the retinopathy of the elderly diabetic frequently possesses numerous exudates (exudative diabetic retinopathy). Cotton-

TABLE 8.1
Some Inherited Disorders that Affect the Eye and Other Structures

Condition	Stored material	Mode of inheritance	Enzymatic defect
A. Mucopolysaccharidoses			
Type I-H (Hurler syndrome)	Dermatan SO_4[a], Heparan SO_4	AR	α-L-Iduronidase
Type I-S (Scheie syndrome, formerly type V)	Dermatan SO_4, Heparan SO_4	AR	α-L-Iduronidase
Type II (Hunter syndrome)	Dermatan SO_4, Heparan SO_4	AR	Sulfoiduronate sulfatase
Type IIIA (Sanfilippo disease A)	Heparan SO_4	AR	Heparan sulfate sulfatase
Type IIIB (Sanfilippo syndrome B)	Heparan SO_4	AR	N-Acetyl-α-D-glucosaminidase
Type IV (Morquio disease)	Chondroitin SO_4, Keratan SO_4	AR	Unknown
Type VI (Maroteaux-Lamy syndrome)	Dermatan SO_4	AR	Unknown
B. Amino acidopathies			
Tyrosinemia		AR	Unknown
Alkaptonuria (ochronosis)		AR	Homogentisic acid oxidase
Albinism		AR	Homogentisic acid oxidase
Oculocutaneous			
Tyrosine-negative oculocutaneous albinism		AR	Tyrosinase
Tyrosine-positive oculocutaneous albinism			
Albinism-hemorrhagic diathesis (Hermansky-Pudlák syndrome)		AR	Unknown
Yellow-type oculocutaneous albinsim		AR	Unknown
Ocular		XR	Unknown

[a] The abbreviations used are: SO_4, sulfate; AD, autosomal dominant; AR, autosomal recessive; XR, x-linked recessive.

Metabolic Disorders 139

TABLE 8.1—Continued

Condition	Stored material	Mode of inheritance	Enzymatic defect
Cystinosis (Lignace-Fanconi syndrome)			
Type I (Infantile)		AR	Unknown
Type II (Juvenile)		AR	Unknown
Type III (Adult)		AR	Unknown
Homocystinuria		AR	Cystathionine synthase
Lowe's oculocerebrorenal syndrome		XR	Unknown
Chorioretinal gyrate atrophy		AR, XR	Unknown
C. **Hemoglobinopathies**			
Sickle cell anemia (Hb S-S disease)		AD	
Sickle cell hemoglobin C disease (Hb S-C disease)		AD	
Sickle cell thalassemia (S-thalassemia)		AD	
D. **Galactosemia**			
Type I. Galactokinase deficiency		AR	Galactokinase
Type II. Galactose-1-phosphate uridyl-transferase deficiency		AR	Galactose-1-phosphate uridyl-transferase
E. **Disorders of Lipid Metabolism**			
I. *Sphingolipidoses*			
G_{M2}-gangliosidosis (type I) (Tay-Sachs disease, infantile amaurotic family idiocy)	G_{M2}-ganglioside + asialo G_{M2}	AR	Hexosaminidase A
G_{M2}-gangliosidosis (type II) (Sandhoff-Jatzhewitz-Pitz disease)	G_{M2}-ganglioside + asialo G_{M2} + globoside G_{A2}	AR	Hexosaminidase A and B
G_{M2}-gangliosidosis (type III) (juvenile G_{M2}-gangliosidosis, Bernheimer-Seitelberger disease)	G_{M2}-ganglioside + asialo G_{M2}	AR	Partial deficiency of hexosaminidase A
G_{M1}-gangliosidosis (type I) (Norman-Landing disease, generalized gangliosidosis)	G_{M1}-ganglioside + asialo-G_{M1} + glycopeptide + sialomucopolysaccharide	AR	β-Galactosidase A, B, and C

Metabolic Disorders

G$_{M1}$-gangliosidosis (type II) (juvenile G$_{M1}$-gangliosidosis)	G$_{M1}$-ganglioside + asialo G$_{M1}$ + glycopeptide + sialo-mucopolysaccharide	AR	β-Galactosidase B and C
Fabry disease (glycolipid lipidosis)	Ceramide trihexoside + ceramide dihexoside	XR	Ceramide-α-trihexosidase (α-galactosidase)
Gaucher disease (cerebroside lipidosis)	Ceramide glucoside	AR	β-Glucosidase
Niemann-Pick disease, infantile form	Sphingomyelin and cholesterol	AR	Sphingomyelinase
Globoid cell leukodystrophy (Krabbe's disease)	Ceramide galactoside (galactocerebroside)	AR	Galactocerebroside β-galactosidase
Farber's lipogranulomatosis	Ceramide + G$_{M3}$-ganglioside	AR	Ceramidase
Metachromatic leukodystrophy, late infantile (sulfatide lipidosis)	Sulfatide	AR	Arylsulfatase A
Metachromatic leukodystrophy, variant	Sulfatide + heparin sulfate	AR	Arylsulfatases A, B, and C
II. *Miscellaneous lipidoses*			
Phytanic acid storage disease (Refsum's disease)	Phytanic Acid	AR	Phytanic acid α-hydroxylase
Neuronal ceroid-lipofuscinosis (Batten-Vogt disease)			
Type I Infantile (Haltia-Stantvuori)	Ceroid-lipofuscin	?	Unknown
Type II Late infantile (Jansky-Bielschowsky)	Ceroid-lipofuscin	?AR	Peroxidase
Type III Juvenile (Spielmeyer-Sjogren-Batten)	Ceroid-lipofuscin	AR	Peroxidase
Type IV Adult (Kufs)	Ceroid-lipofuscin	AD	Peroxidase
Wolman's disease	Cholesterol + triglycerides	AR	Acid lipase
III. *Dyslipoproteinemias*			
Analphalipoproteinemia (Tangier disease)		AR	
Abetalipoproteinemia (Bassen-Kornzweig syndrome)		AR	

142   The Eye

TABLE 8.1—Continued

Condition	Stored material	Mode of inheritance	Enzymatic defect
Hyperlipoproteinemias			
Type I (familial hyperchylomicronemia)		AR	lipoprotein lipase
Type II (hyperbetalipoproteinemia)		AD	3-hydroxy-3 methylglutaryl coenzyme A reductase
Type III (familial hyperbeta and prebetalipoproteinemia)			
Type IV (Carbohydrate induced hyperlipemia)		AD	Unknown
Type V (familial hyperprebetalipoproteinemia)		AD	Unknown
Type VI (familial hyperchylomicronemia with hyperprebetalipoproteinemia)		AR	Unknown
		AR	Unknown
Lecithin-Cholesterol acetyltransferase (LCAT) deficiency		AR	Lecithin-cholesterol acetyltransferase
F. Disorders of Copper metabolism			
Hepatolenticular degeneration (Wilson disease)		AR	Unknown
Menke syndrome (kinky hair disease)		XR	Unknown

TABLE 8.2
Miscellaneous Inherited Conditions Affecting the Eye and Other Structures with Unknown or Incompletely Known Metabolic Defects

Diseases Involving Eye and Other Structures

1. **X-linked diseases**
   Incontinentia pigmenti (Bloch-Sulzberger disease)
   Pelizaeus-Merzbacher disease
   Norrie disease
2. **Autosomal recessive mode of inheritance**
   Rothmund-Thomson (poikiloderma congenita)
   Laurence-Moon syndrome
   Biedl-Bardel syndrome
   Chédiak-Higashi syndrome
   Marinesco-Sjögren syndrome
   Ataxia-telangiectasia (Louis Bar)
   Pseudoxanthoma elasticum
   Xeroderm pigmentosum
3. **Autosomal dominant mode of inheritance**
   Marfan disease
   Paget disease of bone
   von Hipple-Lindau disease
   Primary heredofamilial amyloidosis
   Tuberous sclerosis
   Familial dwarfism with stiff joints
   Sticker's progressive arthrophthalmopathy
   Hidrotic ectodermal dysplasia (Marshall type)
   Myotonic dystrophy
   Familial dysautonomia (Riley-Day syndrome)
4. **Variable mode of inheritance**
   Ehlers-Danlos syndrome (AR, AD)[a]
5. **Mode of inheritance not clear**
   Alport syndrome
6. **To be determined**
   Cornelia de Lange syndrome
   Urbach-Wiethe syndrome (lipid proteinosis)
   Hooft disease

[a] The abbreviations used are: AD, autosomal dominant; AR, autosomal recessive.

146 The Eye

Because other types of vascular disease may coexist with diabetes, the retinopathy also may show changes due to hypertension and arteriosclerosis. Lipemia retinalis may occur in young diabetics with marked acidosis.

In common with several other delayed lesions in diabetes mellitus, the retinopathy is an outcome of vascular disease. Retinal ischemia can account for most features including the cotton-wool spots, capillary closure, microaneurysms, and retinal neovascularization. Retinal ischemia may result from arteriolar narrowing or occlusion, platelet and lipid thrombi in the small vessels, or from atheromatosis of the central retinal artery or ophthalmic artery. The precapillary arterioles are frequently occluded. There is a direct correlation between diabetic retinal microangiography and diabetic glomerular sclerosis.

The retinopathy usually follows 10-20 years after the onset of diabetes. The prevalence of the retinopathy increases with the duration of the disease and 75-90% of patients have retinal changes after 25 years of diabetes. Most individuals with diabetic retinopathy are over the age of 50 years. It is extremely rare for diabetic retinopathy to occur under the age of 10. Women are not only more prone to diabetes than men, but also develop diabetic retinopathy more frequently. The retinopathy does not seem to be closely related to the severity of the diabetes nor to the cause of it. The incidence of diabetic retinopathy is increasing due to the improved life expectancy of treated diabetics and an inability to prevent the retinal disease.

Diabetic Iridopathy

Rubeosis iridis occurs in diabetics with severe retinopathy. It is usually bilateral and is of tremendous clinical importance. Rubeosis iridis is discussed more fully in Chap. 4. Another abnormality of the iris occurs in diabetes. When tissue sections are processed in the usual manner, a lacy vacuolization of the pigment epithelium of the iris, but not of the ciliary body and retina, is sometimes evident (Fig. 8.3). The vacuolization results largely from the loss of glycogen in the preparation of tissue sections. The glycogen storage in the iris reflects an elevated aqueous sugar and is analogous to the Armanni-Epstein phenomenon that occurs with glycosuria in the kidney. Sometimes, proliferation of the pigment epithelium results in several layers of cells. The swollen pigment epithelium cells may rupture, liberating melanin granules which deposit on various structures in the anterior part of the eye. The glycogen storage within the pigment epithelium of the iris is thought to account for the scattering of the iris pigment that can be observed clinically in diabetic irises.

Diabetes and the Lens

A sudden temporary myopia can occur in diabetics and may be the presenting manifestation of the disease. It is caused by an increase in the refractive power of the lens. Hyperglycemia presumably results in an increased sorbitol content of the lens, leading to an imbibition of water and an enlargement of the lens.

At least two distinct types of cataracts are recognized in diabetes mellitus. The true diabetic cataract occurs especially in young diabetics from 15–25 years of age and exceptionally in infancy. The development of this juvenile diabetic cataract is usually directly proportional to the severity of the diabetes. Initially, a blanket of white needle-shaped opacities collects in both lenses just beneath the anterior and posterior lens capsule ("snowflake cataract"). These coalesce until the whole lens becomes opaque. The rate of progression varies with the age of the individual, usually taking a few weeks in adolescents but only days in children. An elevation in blood sugar results in a concomitant formation and accumulation of sorbitol in the lens of young diabetic animals. The so-called senile cataracts are common in elderly diabetics. They occur at

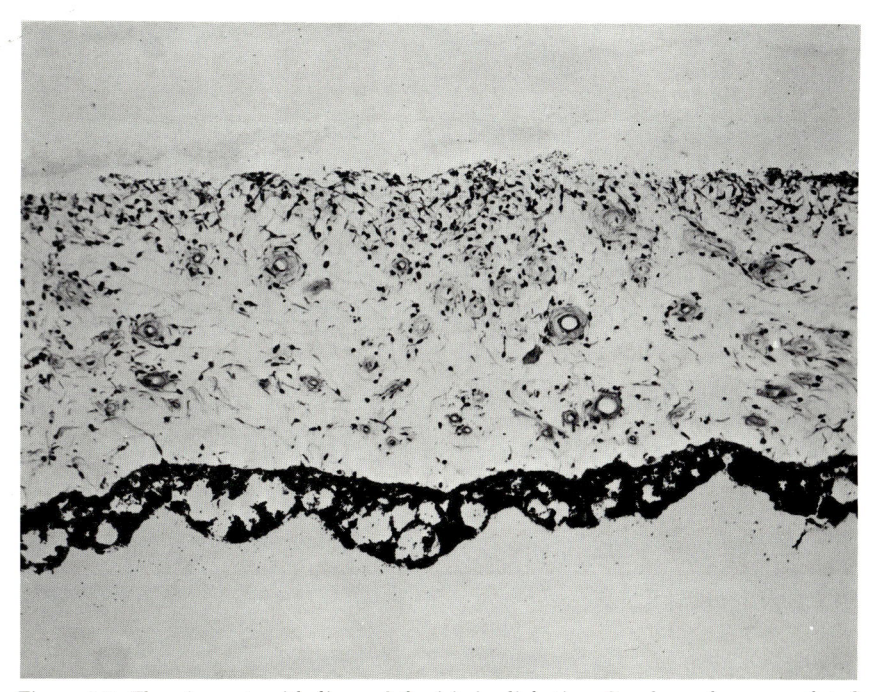

Figure 8.3. The pigment epithelium of the iris in diabetics often has a lacy vacuolated appearance like this after the tissue has been processed in the usual manner. The vacuoles result from the loss of stored glycogen in the preparation of the tissue. (H & E, × 100)

148 *The Eye*

an earlier age than in the general population and progress more rapidly to maturity. Diabetics have an increased incidence of inflammation of the anterior segment of the eye and an increased susceptibility to infections including mucormycosis, a fungal disease that commonly involves the orbit.

Galactosemia

Cataracts are a cardinal feature of two different inherited disturbances in which galactose metabolism is impaired, galactose-1-phosphate uridyltransferase deficiency and galactokinase deficiency (Fig. 8.4). Opacities resembling droplets of oil develop in the center of the lens between the 4th and 5th week of life in a high percentage of untreated cases. By the 10th week of life, but occasionally within 1 week, the lens is usually

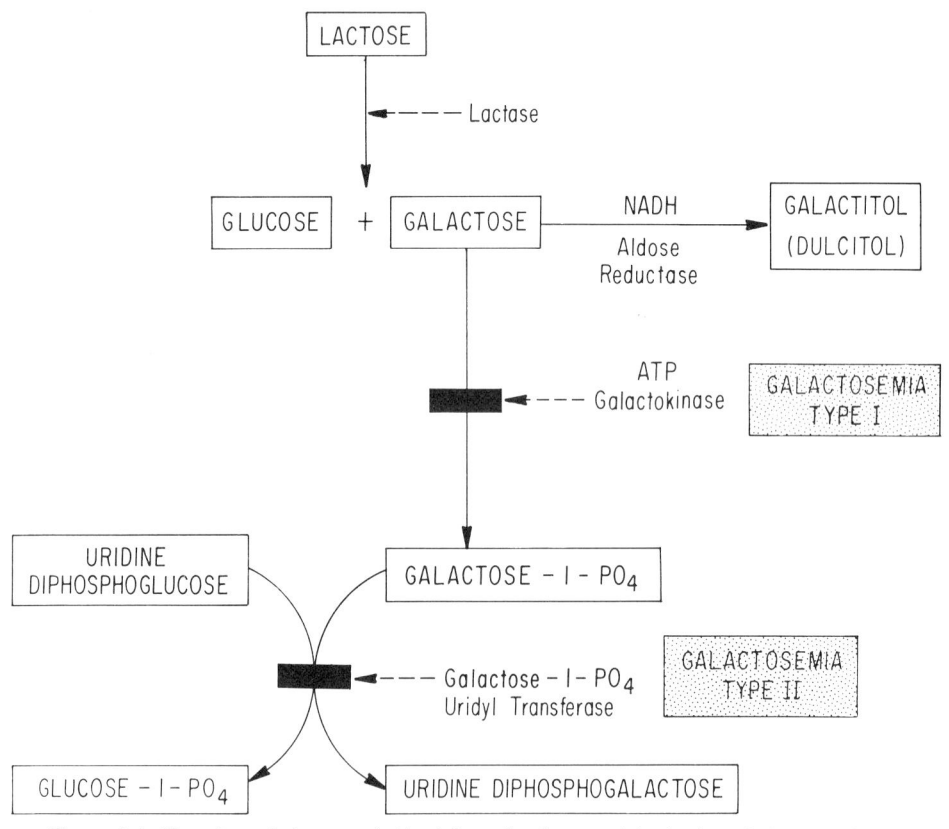

Figure 8.4. The sites of the metabolic defects in the two inherited varieties of galactosemia.

opaque. Galactose is reduced within the lens to dulcitol, which is not metabolized further. The cataract is thought to be due to this accumulation of dulcitol, which remains within the lens as an osmotic force that attracts water. The cataract can be produced experimentally in animals on a high galactose diet. In the young rat, the accumulation of lenticular dulcitol and water is accompanied by progressive opacification of the lens. If incubated lenses are prevented from swelling by counteracting the osmotic pressure of the dulcitol that accumulates within them, the lenses remain clear and retain their capacity to concentrate amino acids from the medium. Analogous cataracts can also be produced by ribose and xylose.

DISORDERS OF MUCOPOLYSACCHARIDES

The cornea and retina are involved in most of the systemic mucopolysaccharidoses. Corneal clouding occurs in all of them. Its occurrence is late in Hunter syndrome (MPS II), and it is rare in the Sanfillippo syndrome (MPS III). Optic atrophy may develop in all systemic mucopolysaccharidoses, presumably as a sequel to the diseased ganglion cells in the retina. A pigmentary retinopathy frequently occurs in Hurler syndrome (MPS I-H), Hunter syndrome (MPS II), Sanfilippo syndromes A and B (MPS III), and Scheie syndrome, (MPS I-S), but not in Morquio syndrome (MPS IV) or Maroteaux-Lamy syndrome (MPS VI).

DISORDERS OF PROTEIN AND AMINO ACID METABOLISM

Alkaptonuria (Ochronosis)

In alkaptonuria, there is an absence of homogentisic acid oxidase, an enzyme which oxidizes homogentisic acid. This is an intermediate compound in the metabolic sequence, whereby phenylalanine and tyrosine are converted to acetoacetic acid which, in turn, is oxidized in the citric acid cycle to yield energy (Fig. 8.5). The inactivity of the enzyme results in an abnormal accumulation of homogentisic acid in the serum. Much is eliminated in the urine, where its presence is diagnostic, even in infancy. The urine is dark or turns dark on standing. Frequently, the condition is detected much later in life, when the progressive deposition of blackish pigment in connective tissues occurs, especially in the cartilage of the ears, nose, and joints. Between the ages of 20 and 30, V-shaped semilunar patches of brownish black scleral pigmentation become clinically evident in both eyes midway between the margin of the cornea and the inner and outer canthi, in the interpalpebral portion of

150   *The Eye*

```
                    ┌─────────────────┐
                    │  PHENYLALANINE  │
                    └────────┬────────┘
                             ↓
                    ┌─────────────────┐
                    │    TYROSINE     │
                    └────────┬────────┘
                             ↓
                    ■ ◄─ ─ ─ Tyrosine Aminotransferase   [TYROSINEMIA]
                             ↓
            ┌────────────────────────────┐
            │ p-HYDROXYPHENYLPYRUVIC ACID │
            └─────────────┬──────────────┘
                          ↓
                          ◄ ─ ─ ─ p-Hydroxyphenylpyruvic acid oxidase
                          ↓
              ┌──────────────────────┐
              │   HOMOGENTISIC ACID  │
              └──────────┬───────────┘
  Polymerized            ↓
              ■ ◄─ ─ ─ Homogentisic acid oxidase   [ALKAPTONURIA]
        ┌──────────────┐ ↓
        │ MELANIN-LIKE │
        │   PIGMENT    │
        └──────────────┘
              ┌──────────────────────┐
              │   MALEYLACETOACETIC  │
              │         ACID         │
              └──────────┬───────────┘
                         ↓
                 Metabolized through
                   Citric Acid Cycle
```

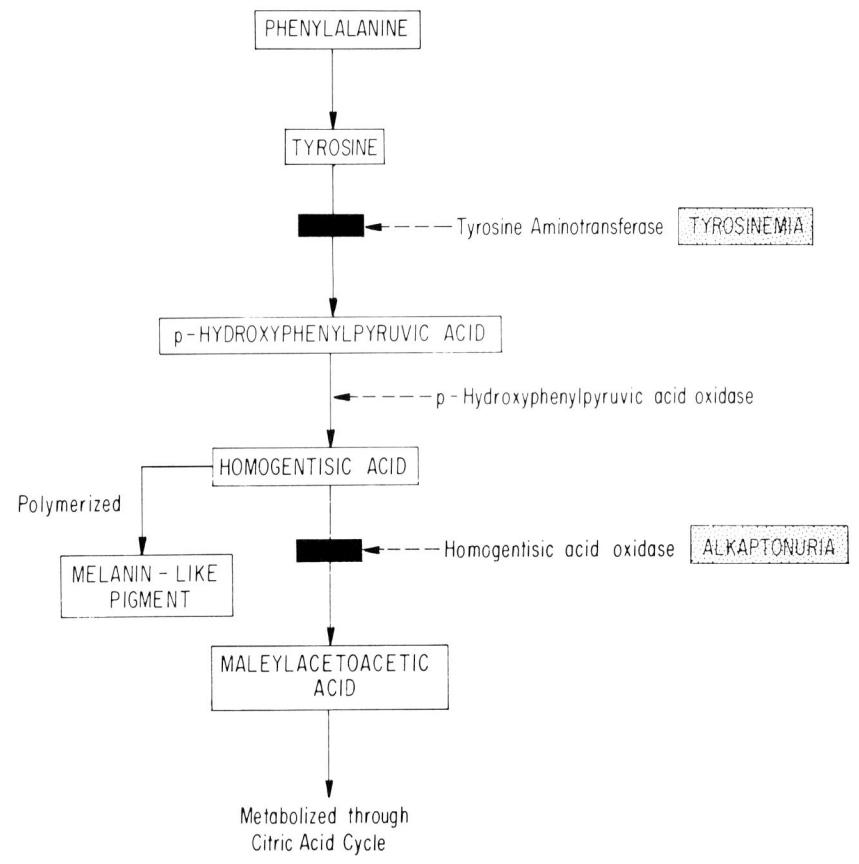

Figure 8.5. Summary of some metabolic pathways involved in the metabolism of phenylalanine and tyrosine.

the eye. The pigment is believed to be a melanin-like polymer of homogentisic acid. A similar scleral pigmentation occurs after prolonged ocular exposure to quinone vapor.

Albinism

There are several different types of albinism with different modes of inheritance. In oculocutaneous albinism, the most severe form of albinism, the skin of the entire body, as well as the hair and eyes, lack pigment. The cutaneous form of albinism does not affect the eyes. Partial albinism occurs in several conditions including the Chédiak-Higashi syndrome. In all varieties of albinism, there is an inability to synthesize the pigment melanin, a metabolite of tyrosine. In one variety of oculocutaneous albinism, the enzyme tyrosinase that mediates two of the metabolic steps between tyrosine and melanin is deficient (Fig. 8.6). Possibilities for the occurrence of other types of albinism include the inhibition of melanin formation itself, or a defect in the mechanism whereby tyrosine is transported from the serum into the melanophores where melanin is synthesized. In albinism, the melanocytes in the skin and uvea are present in normal numbers. They lack mature melanin granules, but contain morphologically normal premelanosomes. The lack of pigment in the choroid makes the choroidal blood vessels visible with the ophthalmoscope. The iris is abnormally translucent, light gray colored, and often hypoplastic. Heterochromia iridis may occur. Melanin is usually present in the pigment epithelia of the retina, ciliary body, and iris, although it is diminished in quantity (Fig. 8.7). The albinotic macula is abnormal, possibly due to the failure of the fovea centralis to develop because of the inability of the retinal pigment epithelium to absorb light. The eyebrows

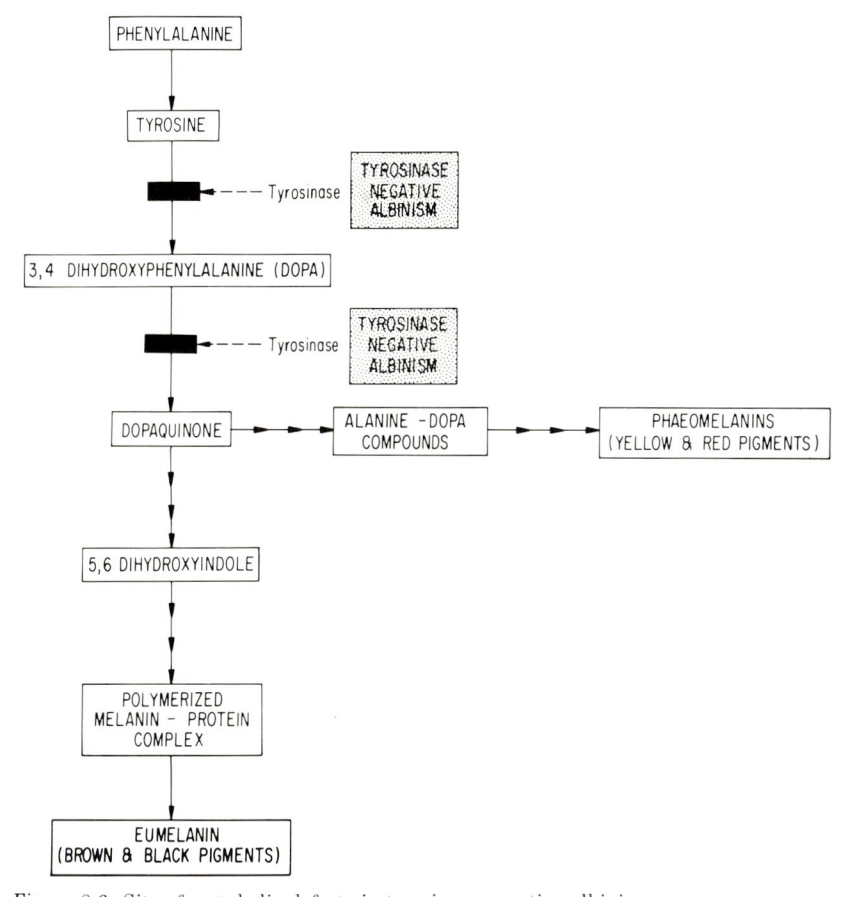

```
PHENYLALANINE
     ↓
  TYROSINE
     ↓
     ■ ←--- Tyrosinase    TYROSINASE
     ↓                     NEGATIVE
                           ALBINISM
3,4 DIHYDROXYPHENYLALANINE (DOPA)
     ↓
     ■ ←--- Tyrosinase    TYROSINASE
     ↓                     NEGATIVE
                           ALBINISM
DOPAQUINONE → ALANINE-DOPA → PHAEOMELANINS
     ↓         COMPOUNDS     (YELLOW & RED PIGMENTS)
5,6 DIHYDROXYINDOLE
     ↓
POLYMERIZED
MELANIN - PROTEIN
COMPLEX
     ↓
EUMELANIN
(BROWN & BLACK PIGMENTS)
```

Figure 8.6. Site of metabolic defects in tyrosinase-negative albinism.

152   The Eye

Figure 8.7. The choroid and outer retina from an individual with partial albinism contains a diminished amount of melanin pigment in the retinal pigment epithelium and uveal melanocytes. (H & E, × 200)

and eyelashes are white. Other ocular abnormalities in albinism include astigmatism, myopia, strabismus, and nystagmus.

Homocystinuria

In homocystinuria, the enzyme cystathionine synthase, needed for the metabolism of homocysteine, is lacking (Fig. 8.8). There is an increased amount of methionine and homocysteine in the serum and an excess of sulfur-containing amino acids in the urine. Ectopia lentis is extremely common in individuals with homocystinuria and is frequently the presenting clinical manifestation. There is a deficiency in the zonules and dislocation of the lens is usually downward and often into the anterior chamber. A forward displacement of the lens may obstruct the pupil and cause glaucoma. Myopia occurs, commonly due to an elongated globe and spherophakia. In a minority of cases, the lens becomes cataractous. There is a high incidence of retinal detachment, even in the absence of cataract surgery. The nonpigmented epithelium of the ciliary body is atrophic and possesses a markedly thickened basement membrane, onto

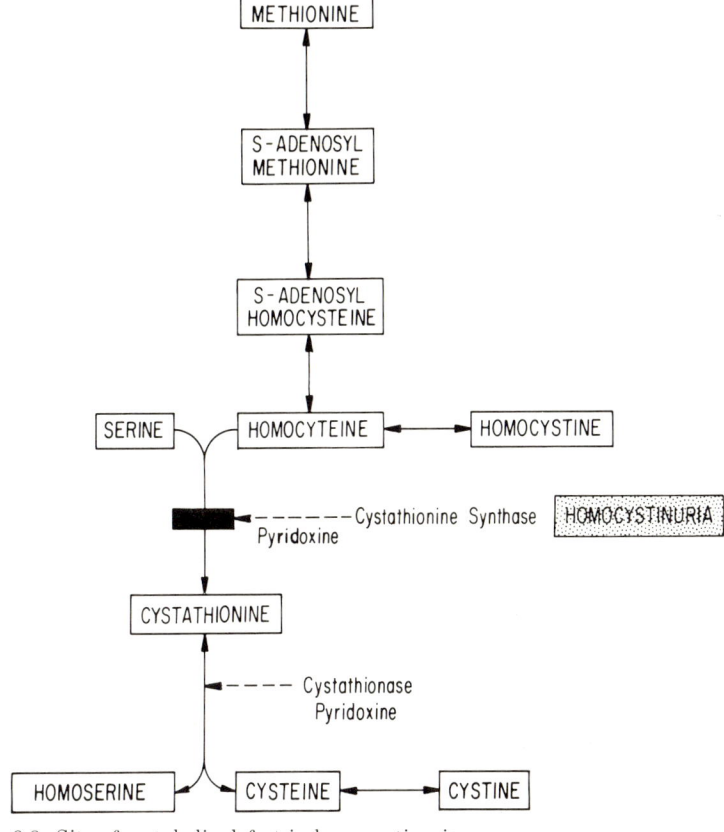

Figure 8.8. Site of metabolic defect in homocystinuria.

which the zonules recoil. Peripheral cystoid degeneration of the retina is marked and is evident at an earlier age than usual. Individuals with homocystinuria have a tendency to vascular thromboses, even at a young age, and the ocular vessels may also be involved. At least three varieties of homocystinuria are now recognized. One type responds to the administration of pyridoxal phosphate.

Lowe's Oculocerebrorenal Syndrome

The manifestations of this inherited disorder include bilateral congenital cataracts (most cases), spherophakia, congenital glaucoma (50% of patients), congenital anomalies of the retina and other ocular structures, mental and physical retardation, aminoaciduria, systemic acidosis, and renal rickets. In heterozygous females, punctate opacities occur in the lens.

Hemoglobinopathies

More than 200 different human hemoglobins have now been characterized. In the genetically determined sickle cell diseases, the hemoglobin molecule has an incorrect amino acid in its protein chain. In hemoglobin S, valine is substituted for glutamic acid in the sixth position of the chain, whereas in hemoglobin C, lysine is substituted in this position. Hypoxic states, such as respiratory infections or high altitude flying, reduce hemoglobin S and C to relatively insoluble long rods which distort the erythrocytes, forming sickle-shaped cells which cannot be molded through the small blood vessels. In the eye, the small peripheral retinal vessels are mainly affected due to a vascular occlusion which results from several mechanisms, including thrombosis. Fluorescein angiography has revealed that the arterioles are the first vessels to become occluded. Tortuosity and dilation of the retinal veins, microaneurysms, and retinitis proliferans, as well as retinal and vitreal hemorrhages occur. The retinopathy resembles that found in diabetes in several respects. However, the peripheral, rather than the postequatorial, vessels are involved in sickle cell disease. Microaneurysms are uncommon in sickle cell disease and are not as conspicuous as in diabetes mellitus. Moreover, in sickle cell disease "dot and blot" hemorrhages are not typical and nor are venous loops around areas of capillary closure. The severity of the retinopathy in the different hemoglobinopathies varies considerably. In hemoglobin S-C disease, the most devastating ocular complications develop. The retinopathy seems to be related to the viscosity of the blood rather than to the duration of the disease. Abnormal blood vessels develop in the midperiphery immediately proximal to the obliterated peripheral vessels. These newly formed vessels grow into the vitreous and bleed, ultimately going on to sensory retinal detachment and leading to visual impairment. Hemorrhagic infarcts can occur in the retina. Neovascularization of the angle can cause glaucoma. Less frequently, central retinal vein thrombosis, iris infarcts, and angioid streaks occur. Vascular stasis may result in peculiar comma-shaped conjunctival vessels, due to afferent and efferent portions of the vascular network being devoid of blood.

DISTURBANCES OF LIPID METABOLISM

Arcus senilis frequently develops during adulthood in hyperlipoproteinemia types II and III (familial hyperbetalipoproteinemia) and less often in hyperlipoproteinemia types I, IV, and V (Figs. 1.6, 1.7). Particularly in the presence of corneal vascularization, lipid may deposit in the cornea in individuals with hyperlipemia (lipid keratopathy). An increase in the triglycerides, pre-β lipoproteins, and chylomicrons in the plasma causes a lactescence of the plasma. This becomes evident clinically in superficial blood vessels and notably in the retinal vessels which possess a salmon pink, creamy white, or grayish color (lipemia retinalis). It occurs if the blood lipid content is more than 5% and the triglyceride level is more than 2000 mg/100 ml. Vision is not affected. The phenomenon occurs most often in young diabetics with marked acidosis.

Aside from xanthelasma (Figs. 1.3, 1.4), intracytoplasmic lipid accumulates in several inherited diseases. In many of these diseases, the stored material is a sphingolipid—a complex lipid containing an unsaturated amino glycol known as sphingosine. When a fatty acid is linked by a peptide bond, the compound is known as ceramide. The sphingolipids include cerebroside, sulfatide, ganglioside, and ceramide polyhexoside. In many of these lipidoses, the metabolic defect can be detected prenatally by amniocentesis. In several lipidoses, the ganglion cells of the retina contain an abundance of stored lipid and a cherry red spot is often evident at the macula (Fig. 1.5). Cherry red spots in the macula may be found in G_{M2}-gangliosidosis, type I (Tay-Sachs disease), G_{M2}-gangliosidosis, type II (Sandhoff disease), G_{M1}-gangliosidosis (generalized gangliosidosis), Niemann-Pick disease, Farber's lipogranulomatosis, metachromatic leukodystrophy, mucolipidosis I (lipomucopolysaccharidosis), and sea blue histiocyte syndrome (chronic Niemann-Pick disease), as well as other lipidoses. A pigmentary retinopathy occurs in neuronal ceroid lipofuscinosis.

G_{M2}-gangliosidosis (Tay-Sachs Disease)

This sphingolipidosis is characterized by the accumulation within the neurons of the brain and retina of G_{M2}-ganglioside. The condition has a high gene frequency in Ashkenazi Jews. The primary enzymatic defect appears to be an absence of β-D-N-acetyl hexosaminidase A which cleaves the terminal N-acetyl galactosamine residue from G_{M2}-ganglioside. A cherry red spot occurs at the macula.

Glycolipid Lipidosis (Fabry's Disease)

In Fabry's disease, cerebroside trihexoside accumulates in various tissues, including the eye. Severe manifestations in this X-linked disease occur in hemizygous males, whereas heterozygous females are mildly affected. The disease is of particular importance because characteristic whorl-like opacities in the cornea can be readily detected with the slit lamp not only in the severely affected male, but also in the heterozygous female. The small blood vessels of the eye and other tissues are dilated and tortuous. Posterior capsule lens opacities with spoke-like radiations are common.

Phytanic Acid Storage Disease (Refsum's Disease)

A pigmentary retinopathy and cataracts occur in a disorder in which there is an inability to oxidatively metabolize phytanic acid (3,7,11,15-tetramethylhexadecanoic acid), a fatty acid of dietary origin.

DISORDERS OF HEAVY METALS

Hypercalcemia

Calcium salts may deposit in the conjunctiva and occasionally in the cornea in hypercalcemia. The latter results from many conditions. These include hypophosphatasia, sarcoidosis, and hyperparathyroidism, as well as the excessive ingestion of vitamin D or calcium carbonate and milk (milk-alkali syndrome).

Hypocalcemia

In hypocalcemia, bilateral punctate and subcapsular opacities sometimes develop in the lens, particularly beneath the posterior capsule, if the serum calcium falls to a sufficiently low level to cause tetany also. They are not reversible when the serum calcium returns to normal.

Hepatolenticular Degeneration (Wilson's Disease)

A decreased synthesis of structurally normal ceruloplasmin and an increased serum-unbound copper are cardinal to this disease. Copper deposits in various tissues, including the eye. A greenish brown deposition of copper occurs in the periphery of Descemet's membrane (Fig. 8.9). This accounts for the frequently observed Kayser-Fleischer ring, which usually is most marked in the upper and lower margins of the cornea where it often begins. Copper also may accumulate in the capsule of the lens (sunflower cataract). In patients with a retained copper or bronze intraocular foreign body, copper deposits in similar locations.

MISCELLANEOUS INHERITED DISEASES WITH UNKNOWN METABOLIC DEFECT

Marfan's Syndrome

Bilateral ectopia lentis occurs in about 80% of individuals with Marfan's syndrome. The lenses are small and spherical and are most often displaced posteriorly (Fig. 8.10). The dilator pupillae is hypoplas-

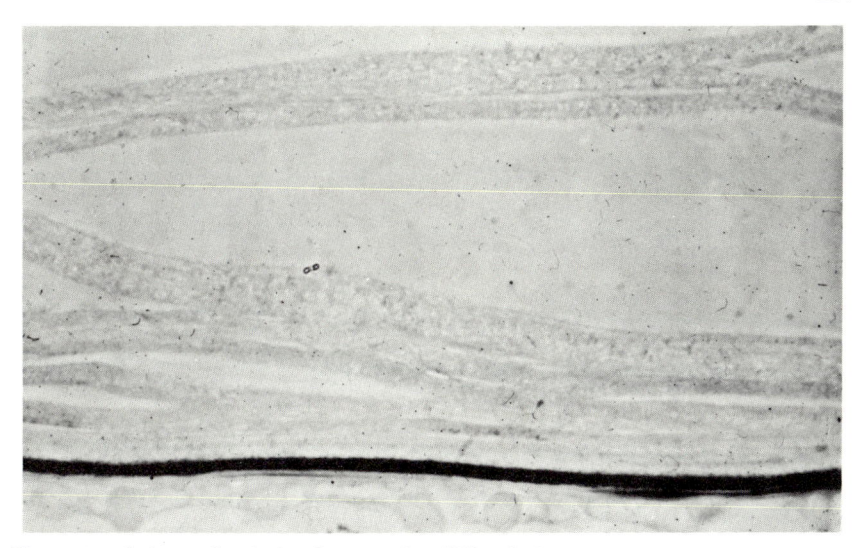

Figure 8.9. In hepatolenticular degeneration (Wilson's disease), copper deposits in Descemet's membrane. This accounts for the Kayser-Fleischer ring. (Unstained, × 640)

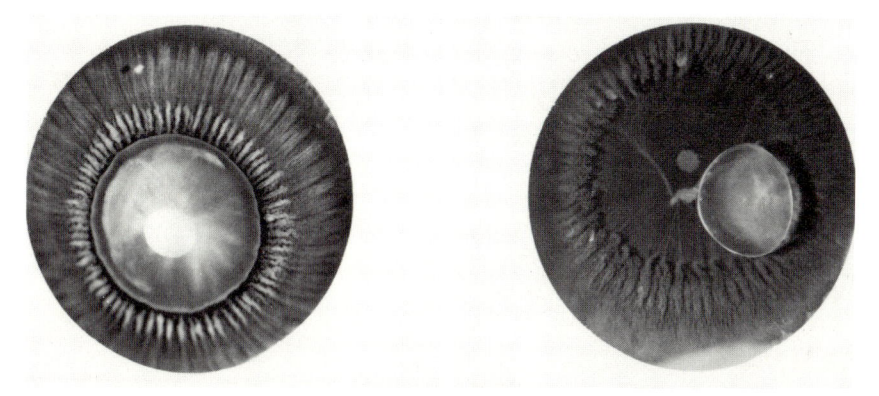

Figure 8.10. The crystalline lens, ciliary body, and iris are viewed from behind. In contrast to the normal eye (*left*), the lens in Marfan's syndrome is small, spherical, and displaced. The pupil in the abnormal eye is also miotic.

tic, making the pupil difficult to dilate. Glaucoma frequency develops due to an anomalous trabecular meshwork or to the sequelae of a dislocated lens. High myopia and retinal detachment are common.

von Recklinghausen's Disease of Nerves

In von Recklinghausen's disease of nerves, neurofibromas, schwannomas, and other neoplasms, including meningiomas and optic nerve

158   *The Eye*

gliomas, are common (Fig. 8.11). Neurofibromas may involve the innervation of the eye, eyelids, and orbit. Involvement of ciliary nerves thickens the uvea and may be accompanied by the presence of ganglion cells in the choroid. In the eyelid or orbit, a diffuse or beaded thickening of nerves and their branches may occur (plexiform neurofibroma). Variable sized semiglobular or pedunculated masses may project from the skin of the eyelid or elsewhere (mollusca fibrosa). Congenital glaucoma is common. A defect in the orbital bones with a posterior orbital encephalocele and pulsating exophthalmos can occur. Hamartomas may occur in the uvea, retina, and optic nerve. Several oval light brown macules (café-au-lait spots) frequently exist in the skin and can involve the eyelids.

von Hippel-Lindau Disease

Yellowish or red, rounded angioblastic hemangiomas, fed by abnormal tortuous vessels, may develop, especially in the peripheral retina of one or both eyes. Progressive retinal hemorrhage and exudation may result in retinal scarring. Macular exudates can decrease central vision. The lesions may involve the choroid and sclera. Cerebellar hemangioblastomas, hypernephromas, and congenital cysts in the pancreas, adrenals, and kidneys are common in this disorder.

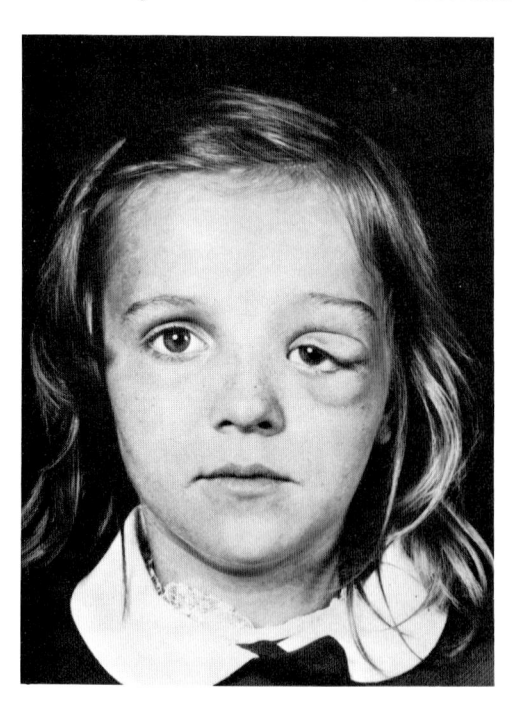

Figure 8.11. von Recklinghausen's disease of nerves with neurofibromas of the upper and lower lids of the left eye.

Tuberous Sclerosis (Bourneville's Disease)

One or more small, grayish white masses, consisting of densely packed glial cells, occur in the retina in about 4% of patients with tuberous sclerosis. These retinal phakomas (retinal gliomas) frequently calcify.

Norrie's Disease

Norrie's disease presents with blindness and a vascularized retrolental mass at birth or during early infancy. The retina is abnormal, detached, and markedly disorganized. In the course of time, the eyes usually become phthisic. Cerebral lesions frequently cause dementia and psychosis. A sensory hearing loss is common.

Inherited Retinal Diseases

Some inherited degenerative diseases of the retina affect the central macula, but in others, involvement of the peripheral retina is the initial or most prominent finding. Among the more common hereditary degenerative diseases involving the macula, and hence affecting central vision primarily, are Stargardt's disease, Best's disease, and Doyne's honeycomb choroiditis. Histologic examination of eyes with these diseases have been rare and not performed at a sufficiently early stage to establish the primary morphologic lesion.

Hereditary Color Blindness

The existence of three types of cones, each with a different pigment (erythrolabe (red-sensitive pigment), chlorolabe (green-sensitive pigment), and cyanolabe (blue-sensitive pigment)), has been demonstrated by studying the absorption spectrum of the retinal outer segments, after passing a microbeam of light through them. In living individuals, the reflected light can be studied with a retinal densitometer and it is possible to deduce the changed pigment density of various wavelengths to bleaching and regeneration. One or more visual pigment is presumed to be absent or deficient in inherited color blindness. Individuals may lack the ability to distinguish one color (dichromats). Some patients are not able to distinguish red from green. Two varieties of red-green color defects exist. Individuals may be insensitive to red light (protanopes), because they lack the red-sensitive pigment. Other individuals (deuteranopes) see red as brightly as normal but find it the same color as green. They are believed to lack chlorolabe. There is no part of the visual spectrum to which deuteranopes are insensitive. Cone monochromatism is a rare abnormality in which the cones appear to possess only cyanolabe.

OXYGEN TOXICITY

Retinopathy of Prematurity (Retrolental Fibroplasia)

The retinopathy of prematurity was one of the leading causes of blindness of infants in the United States and other countries between about 1940 and the early 1950s. The condition is almost restricted to premature infants and occurs in eyes that are, as a rule, normal sized or slightly smaller than usual. The peripheral retina, especially on the temporal side, does not normally become vascularized until about the end of fetal life. Should a premature infant be exposed to oxygen in sufficient amounts, the growing blood vessels in the immature retina obliterate and the peripheral retina fails to vascularize. The more mature the retina, the less the vaso-obliterative effect of hyperoxia. A subsequent, intense proliferation of vascular endothelium and fibroblasts commences on cessation of exposure to oxygen as a secondary response to retinal ischemia (Fig. 8.12). This neovascularization begins at the junction of the avascular and vascularized portions of the retina and becomes apparent by about 5–10 weeks after removal from the incubator. Sometimes, neovascularization of the iris occurs. The retinopathy

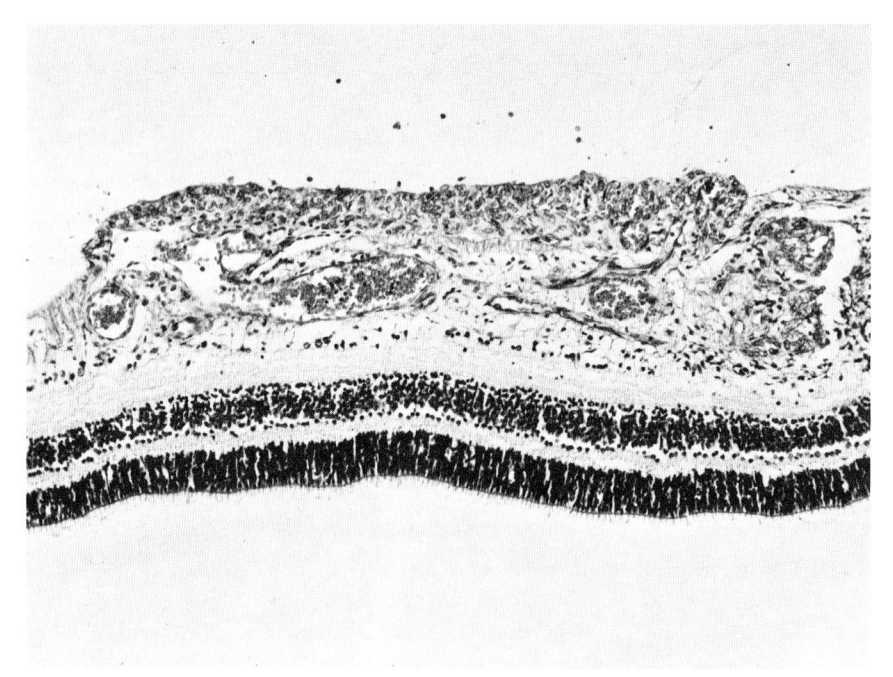

Figure 8.12. The pronounced vascularization of the inner portion of this retina followed the cessation of oxygen administration to a premature infant. (H & E, × 100)

progresses in approximately 25% of cases to a cicatricial phase which culminates in a retrolental, fibrovascular mass which incorporates a detached retina (Fig. 8.13). When this stage develops, it occurs by about 3-6 months.

The pathogenesis of the retinopathy of prematurity has been extensively studied experimentally. In susceptible animals, hyperoxia for several minutes results in a constriction of the retinal arterioles, followed by the disappearance of the capillary circulation. Later, the vessels reopen and remain patent for about 6 hours, after which time they again gradually close. Continuous exposure to high levels of inspired oxygen for prolonged periods of time gives rise to permanent closure.

VITAMIN DEFICIENCIES AND EXCESSES

Vitamin deficiencies are still common in many primitive communities and in the developing parts of the world where the diet is inadequate. In diseases like sprue, celiac disease, and mucoviscidosis, vitamins may not be absorbed. A wide variety of ocular abnormalities result from vitamin deficiencies in experimental animals. This is felt to be almost certainly true in man. It is, however, more difficult to prove because most human vitamin deficiencies are associated with deficiencies of other essential nutrients.

The fat-soluble vitamin A is derived from carotene which is present in many plants. β-Carotene is converted in the intestine to retinol (vitamin A alcohol) and its ester. Vitamin A, which has an essential role in the visual purple (rhodopsin), is stored in the liver. Rhodopsin, the visual pigment of rods, is pink and bleaches to a pale yellow in light. It regenerates in the dark when in contact with the retinal pigment epithelium. One of the earliest signs of vitamin A deficiency is the failure

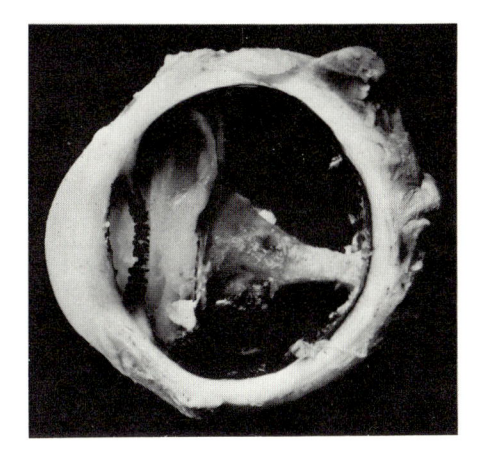

Figure 8.13. The retina is totally detached and adherent to a fibrovascular mass behind the lens in this advanced stage of the retinopathy of prematurity (retrolental fibroplasia).

of formation of rhodopsin from retinene (vitamin A aldehyde) and a protein opsin. This results in an increase in the time required for dark adaptation, a disappearance of the outer segments of the photoreceptors, myelin figures in the retinal pigment epithelium, and eventually night blindness (nyctalopia). A deficiency of vitamin A also results in a loss of the integrity of epithelial structures. The ducts of the lacrimal gland become obstructed as a result of keratinization. Small, silvery gray plaques, with a foamy surface, occur in the superficial exposed bulbar conjunctiva, especially in male children. These are usually in the interpalpebral fissure on the temporal side of the globe (Bitot's spots). The foamy appearance is thought to be due to the presence of *Corynebacterium xerosis*. Corneal keratinization, associated with an inflammatory infiltrate, occurs. Corneal clouding and softening (keratomalacia) characteristically occur in infants with an associated protein deficiency. It is usually bilateral and is followed by corneal ulceration.

Vitamin A, in excess, can produce papilledema secondary to increased intracranial pressure. Band keratopathy with calcium deposition in the cornea and conjunctiva may follow prolonged vitamin D ingestion. High doses of nicotinic acid have been reported to produce cystoid macular edema.

ENDOCRINE DISORDERS

Hyperthyroidism

In hyperthyroidism, exophthalmos of variable severity is common and may precede or succeed the other manifestations of the disease (Fig. 8.14). It is usually bilateral, although one side may be involved earlier or more extensively than the other. Mild exophthalmos is common in early adult life, especially in females. It may be severe and progressive, particularly in middle life, when the exophthalmos no longer correlates well with the state of the thyroid function.

The exophthalmos seems to be largely due to an increase in orbital water imbibed by the osmotic pressure of mucopolysaccharides. In slight cases, there is no overt abnormality on gross or microscopic examination of the orbital tissue. In severe cases, the extraocular muscles are often markedly swollen. There may be a lymphocytic infiltration in the muscle. Some patients with severe exophthalmos also have pretibial or localized myxedema, reflecting a similar disease process.

In patients with Graves' disease the exophthalmos has been attributed to the presence of an antibody. This is a component of the 7 S γ-globulin moiety of serum protein and has been designated long acting thyroid stimulator. This view has been questioned because many patients with exophthalmos due to hyperthyroidism do not demonstrate long acting thyroid stimulator activity. The removal of the thyroid enhances the degree of and/or the incidence of hormone-induced exophthalmos. Lid retraction in hyperthyroidism does not parallel the increase in thyroxin that potentiates the action of epinephrine.

Adrenal Insufficiency

Chronic hypofunction of the adrenal gland (Addison's disease) is accompanied by an increased pigmentation of the conjunctiva and skin. This probably results from the loss of the inhibitory action of the adrenal gland upon the pituitary to oppose production of melanin-stimulating hormone.

DRUGS AND TOXINS

Several commonly used drugs can cause untoward reactions due to prolonged administration, incorrect dosage, or sensitivity reactions. Ethambutol and isoniazid can produce optic neuritis, the latter secondary to a pyridoxine deficiency. The prolonged administration of chloroquine, an antimalarial agent used in the treatment of various rheumatoid and connective tissue disorders, and, less often, hydroxychloroquine, can produce mild corneal opacities, severe retinal pigmentary changes and decreased vision. Chlorpromazine can result in pigmentation of the skin and conjunctiva and cornea in areas exposed to the sun. Marked

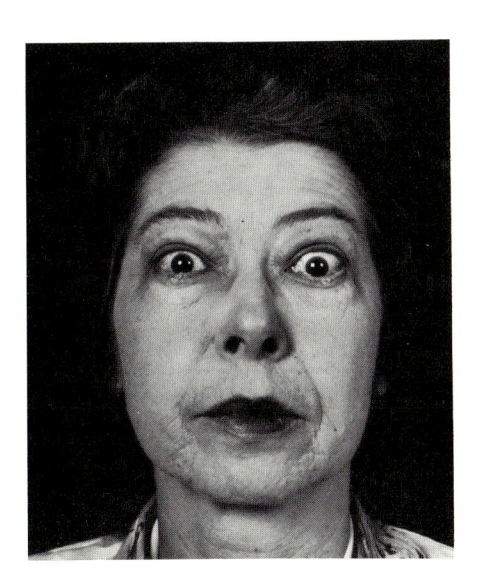

Figure 8.14. Exophthalmos secondary to hyperthyroidism. There is bilateral lid retraction and exophthalmos of the left eye.

retinal pigmentary changes and decreased vision may follow thioridazine (Mellaril) therapy. Anticholinergic drugs can precipitate attacks of narrow angle glaucomas in predisposed eyes. Digitalis intoxication often is manifested with a blurred vision and disturbed color perception. Pigmentation of the conjunctiva, cornea, or eyelids may follow prolonged topical instillation of epinephrine used in the therapy of glaucoma. Cataracts can result from a wide variety of orally or topically administered drugs. These include triparanol (Mer-29) and corticosteroids. Corticosteroids, in addition to causing posterior subcapsular cataracts, can produce increased intraocular pressure. Some individuals on quinine develop a reversible constriction of the retinal arterioles with pallor of the disc, constriction of the visual fields, and a cherry red spot at the macula as seen in central retinal artery occlusion.

Aside from drugs, many ingested chemicals have an adverse effect on the eye. Methanol is but one example. Patients who survive methanol poisoning frequently develop optic atrophy.

REFERENCES

ANDERSON, D. R. Mechanism of Graves' disease and endocrine exophthalmos. Am. J. Ophthalmol. 68:46–57, 1969.

ASHTON, N. Some aspects of the comparative pathology of oxygen toxicity in the retina. Br. J. Ophthalmol. 52:505–531, 1968.

CAIRD, F. E., PIRIE, A., AND RAMSELL, T. G. Diabetes and the Eye, Blackwell Scientific Publications, Oxford and Edinburgh, 1969.

COGAN, D. G., AND KUWABARA, T. The sphingolipidoses and the eye. Arch. Ophthalmol. 79:439–452, 1968.

DEUTMAN, A. F. Hereditary Dystrophies of the Posterior Pole of the Eye. Charles C Thomas, Assen, Netherlands, 1971.

FONT, R. L., AND FINE, B. S. Ocular pathology in Fabry's disease. Am. J. Ophthalmol. 73:419–430, 1972.

GARNER, A. Pathology of diabetic retinopathy. Br. Med. Bull. 26:137–142, 1970.

GOLDBERG, M. F. (Ed). Genetic and Metabolic Eye Disease. Little, Brown & Co., Boston, 1974.

GRANT, W. M. Toxicology of the Eye, 2nd edition. Charles C Thomas, Springfield, Ill., 1974.

HENKIND, P., AND ASHTON, N. Ocular pathology in homocystinuria. Trans. Ophthalmol. Soc. U.K. 85:21–38, 1965.

HOEPNER, J., AND YANOFF, M. Ocular anomalies in trisomy 13–15. Am. J. Ophthalmol. 74:729–737, 1972.

KINOSHITA, J. H. Cataracts in galactosemia. The Jonas S. Friedenwald Memorial Lecture. Invest. Ophthalmol. 4:786–799, 1965.

KRILL, A. E. Principles of genetics. *In* Hereditary Retinal and Choroidal Diseases, Vol. 1, Chap. 1, pp. 1–72. Harper & Row, Publishers, New York, 1972.

KUWABARA, T., KINOSHITA, J. H., AND COGAN, D. G. Electron microscopic study of galactose-induced cataract. Invest. Ophthalmol. 8:133–149, 1969.

McKUSICK, V. L. Mendelian Inheritance in Man; Catalogs of Autosomal Dominant, Autosomal Recessive, and X-Linked Phenotypes, 3rd edition. John Hopkins Press, Baltimore, 1971.

McKusick, V. A. Heritable Disorders of Connective Tissue. 4th edition, C. V. Mosby, Co., St. Louis, 1972.

Moses, R. A. (Ed). Adler's Physiology of the Eye, 5th edition. C. V. Mosby, Co., St. Louis, 1970.

Potts, A. M. (Ed). The Assessment of Visual Function, Chaps. 5 and 6. C. V. Mosby, Co., St. Louis, 1972.

Stanbury, J. B., Wyngaarden, J. B., and Fredericksen, D. S. (Eds). The Metabolic Basis of Inherited Disease, 3rd edition, McGraw-Hill. New York, 1972.

Waardenburg, P. J., Franceschetti, A., and Klein, D. Genetics and Ophthalmology (2 volumes). Royal Van Gorcum. Publisher Assen, Netherlands, 1961.

Wald, G. Molecular basis of visual excitation. Science 162:230–239, 1968.

9

PHYSICAL AND CHEMICAL INJURIES

The eye can become injured in a wide variety of ways *in utero*, during birth, and throughout the life span of the individual. Most ocular injuries are accidentally or maliciously produced. Rarely are some self-inflicted. Ocular injuries sustained in play, sport, physical assault, automobile accidents, and certain occupations are particularly common.

DIRECT PHYSICAL TRAUMA TO THE EYE

Direct blows to the eye are most frequently directed toward the less protected temporal side of the eye. The effects of physical trauma to the eye vary with the nature of the injury. A contusion of the globe may disrupt the corneoscleral envelope at the point of impact, on the opposite side (contrecoup), or at the site of an inherent weakness such as equatorial sclera, corneoscleral limbus, or a staphyloma. Rupture of the globe commonly results in loss of intraocular contents and hemorrhage. Conjunctival lacerations frequently involve the underlying eyelid or globe. Injuries to the intraocular structures may occur in the absence of surface wounds.

Perforating wounds of the globe, such as lacerations of the corneoscleral envelope, are often complicated by loss of aqueous humor and prolapse or incarceration of uveal tissue, lens, and/or vitreous. There may be a loss of tissue at the wound.

With corneal punctures and other accidental or surgical perforating ocular wounds, minute fragments of the corneal or conjunctival epithelium may become implanted into the anterior segments of the eye. The displaced cells may survive and line a thin walled cyst (epithelial implantation cyst). These contain clear fluid and sediment. They may occur in the anterior chamber, iris, conjunctiva, and eyelids. Particularly when the wound edges are inadequately apposed, the surface epithelium may migrate into the anterior chamber and line it ("epithelial ingrowth"). There may be concurrent damage to many intraocular structures. Hemorrhage commonly follows ocular injury, particularly in older individuals, and may be located in the eyelid, conjunctiva, or anterior chamber, as well as in other intraocular locations. It may follow lacerations that extend across blood vessels.

Cornea

Direct trauma to the eye commonly abrades the corneal epithelium producing defects that are readily demonstrated in the living patient, after staining the surface of the eye with fluorescein. The corneal epithelium readily regenerates and cells migrate across the defect from the margin and usually heal the abrasion within 1 day or 2, although local anesthetics delay epithelialization. Occasionally, recurrent corneal erosions ensue. Contact lenses, especially if worn for prolonged periods of time, commonly cause corneal abrasions. Subepithelial corneal edema (Sattler's veil) may occur when the cornea is separated from the atmosphere by contact lenses. Corneal wounds that extend into the stroma heal by scar tissue and the cornea does not reach its maximum tensile strength until several months have passed. Bowman's zone does not reform after it has been damaged. There may be delayed healing or an incomplete apposition of the wound edges. Granulation tissue may form at the site of a nonapproximated defect.

After a corneal contusion, Descemet's membrane may rupture and cause an edematous cornea. The torn Descemet's membrane fails to reunite and becomes retracted and coiled on itself (Descemet's scroll). Endothelial cells cover the defect on the inner cornea and lay down a thin, newly formed Descemet's membrane.

Iris

The iris may be forced backward at its insertion (recession of the iris) and result in a deeper and more rounded angle than normal. If the zonular fibers are ruptured, the iris becomes retroverted against the ciliary body (inversion of the iris). Permanent mydriasis is common after iridial contusions. A contusion may tear the iris radially through the pupil border, sever the sphincter pupillae (rupture of the sphincter pupillae), and even disinsert the iris root from the ciliary body (iridodialysis). Lacerations of the iris may also follow perforating or penetrating ocular injuries. Tears in the iris do not heal.

Ciliary Body

After an ocular contusion, the ciliary body may be torn from the scleral spur (cyclodialysis). A contusion of the ciliary body may tear the ciliary muscles and processes. Damage to the ciliary muscle or its innervation may lead to cycloplegia. Occasionally, hypotony follows diminished aqueous secretion. If the anterior aspect of the ciliary body is torn, the major vascular ring of the iris or the anterior ciliary vessels are severed and massive intraocular hemorrhage into the anterior and posterior chambers, suprachoroidal space, and vitreous may result.

Choroid

A severe contusion of the anterior segment of the eye often results in single or multiple choroidal ruptures. These occur especially in the posterior pole, where the choroid is normally anchored to the sclera by the short posterior ciliary arteries. These tears are concentric with the optic disc and are most frequent on the temporal side of the eye. They can result in hemorrhage within the choroid, suprachoroid, and subretinal spaces, if Bruch's membrane is ruptured.

Retina

A contusion of the eye may result in edema and scattered contrecoup retinal hemorrhages near the posterior pole ("comotio retinae"). It often resolves without permanent damage. Macular edema usually subsides without permanent damage in 3-4 weeks. The clear cystoid spaces in the outer plexiform layer may coalesce and rupture, causing a "macular hole." Small retinal lacerations ("retinal tears") may follow an ocular contusion. These are usually single, frequently horseshoe-shaped, and involve especially the superior nasal quadrant of the eye (Fig. 9.1). Extensive lacerations of the retina are generally associated with injuries to other parts of the eye. The retina may disinsert at the ora serrata.

Lens

Blunt trauma may rupture the zonular fibers and dislocate the lens into the vitreous or anterior chamber. A variety of cataracts may follow ocular injury. A contusion of the lens may provoke proliferation of the anterior subcapsular epithelium (anterior subcapsular cataract). A forcible separation of the lens fibers from the suture lines can cause a rosette-shaped cataract. Perforations of the lens capsule permit aqueous to enter and swell the lens fibers, producing an opaque lens. A swollen lens may push the iris against the cornea, closing the filtration angle and producing a secondary glaucoma.

Eyelids

Lacerations of the upper eyelid can involve the levator palpebrae superioris and superior rectus muscles. Lacerations of the lower lid commonly occur in the medial portion, severing the inferior lacrimal

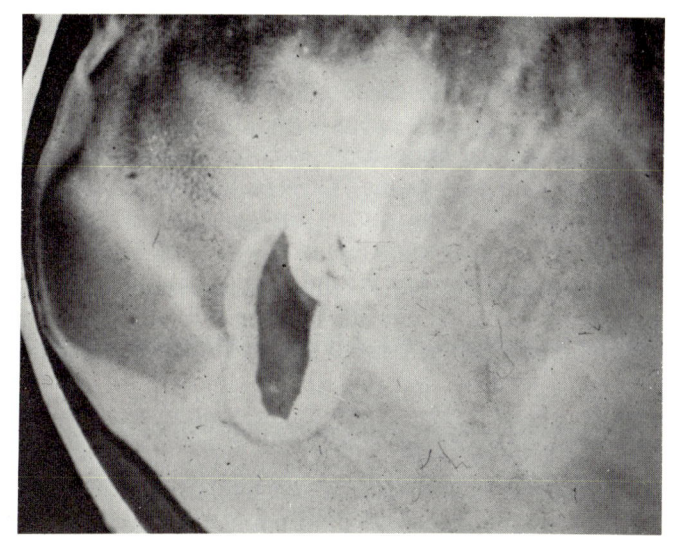

Figure 9.1. Retinal tear in an autopsy specimen. The tear is located in the peripheral retina near the ora serrata. An elevated flap is present with rolled edges of retina around the open break.

duct, which drains the tears toward the lacrimal sac and nasolacrimal duct.

Orbit

Blast injuries and blunt injuries to the eye, such as those in automobile accidents, commonly fracture the orbital bones extensively ("blow-out fracture"). A markedly elevated intraorbital pressure depresses the paper-thin floor of the orbit, prolapsing the orbital contents into the maxillary sinus. Fractures of the superior orbital rim may involve the trochlea of the superior oblique muscle. Fractures or perforating injuries of the medial wall of the orbit permit air from the paranasal sinuses to enter the orbit (orbital emphysema) and cutaneous tissue of the lid, especially during violent blowing of the nose. Structures such as the nasolacrimal ducts, which pass through the bony canals, may be severed.

Fractures of the orbital rim most commonly involve the area of the zygoma and maxilla. Penetrating injuries of the orbit may perforate the posterior walls of the orbit and enter the brain or sinuses. Orbital edema or hemorrhage into the orbit can result in proptosis.

SURGICAL WOUNDS

Surgical procedures constitute a specific type of ocular injury. Incised surgical wounds, as in full thickness corneal grafts or cataract extrac-

172 The Eye

Figure 9.3. Ferruginous intraocular foreign bodies provoke a marked retinal degeneration which is often associated with numerous macrophages that contain iron. (H & E, × 180)

periphery of Descemet's membrane, the adjacent corneal stroma, the anterior lens capsule, and the peripheral lens fibers ("sunflower cataract") notably are affected.

Foreign bodies embedded in the cornea, such as those composed of iron and copper, can discolor the tissue. Should a foreign body become impacted in the lens and the capsular wound heal, a minimal reaction usually results even if the foreign body is copper or iron.

Eyelashes are occasionally carried into the eye during an accidental or surgical perforating wound. They may adhere to the iris and remain in the anterior chamber without any evident reaction or may provoke a foreign body reaction.

The eye may be adversely affected not only by the accidental exposure to foreign materials, but by materials used in the therapy of certain ocular disorders.

ELECTROMAGNETIC RADIATION

The ocular tissue is particularly susceptible to the adverse effect of electromagnetic radiation (Fig. 9.4). This is largely due to the superficial location of the eye and the focusing mechanisms which permit the radiant energy, such as that produced by short infrared waves and visible

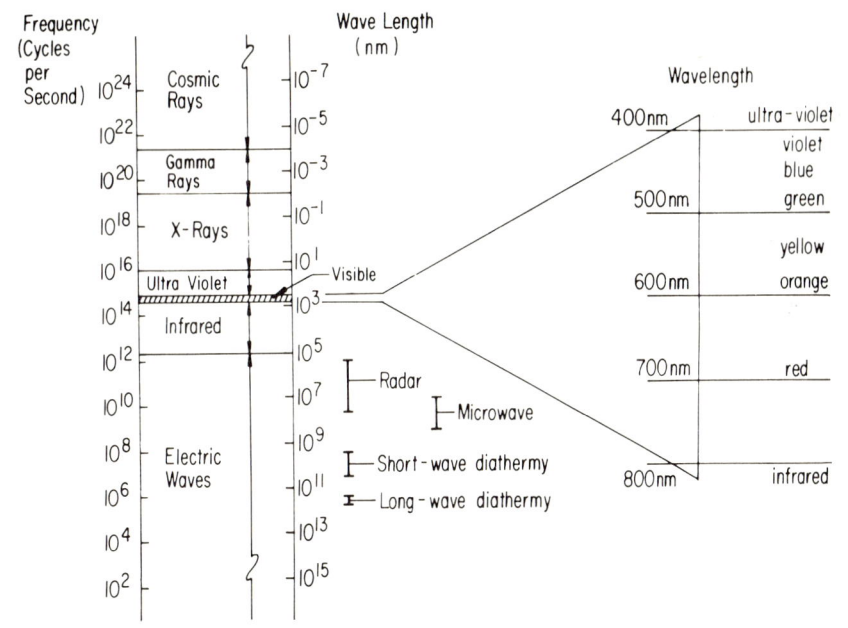

Figure 9.4. The electromagnetic spectrum.

light, to be brought to a focal point on the retina where severe retinal damage may result. The eye is the part of the body most easily damaged by visible light because the focusing mechanism increases the energy sensitivity nearly 100,000 times.

The amount of light entering the eye is influenced by the degree of lid closure and the area of the pupil. Some protection against injury by visible, infrared, and ultraviolet light is provided by blinking of the eye. An additional defense is the normal constriction of the pupils to visible, but not invisible, light.

Radiant energy from the sun, nuclear explosions, intense visible light, lasers, and a variety of other sources can injure the eye. The tissue is not affected by the radiant energy which is reflected or transmitted through it, but only by that which is absorbed.

Flashes, as in explosions, cause an enormous amount of infrared, visible, and ultraviolet radiant energy to be released. Some radiant energy, such as visible light, infrared, and other electromagnetic waves, is degraded into heat and causes burns due to an increase in temperature. Should the temperature become sufficiently elevated, the cellular proteins coagulate.

The amount of radiation transmitted through or absorbed by the ocular tissues varies with the wavelength of the radiation. The longest

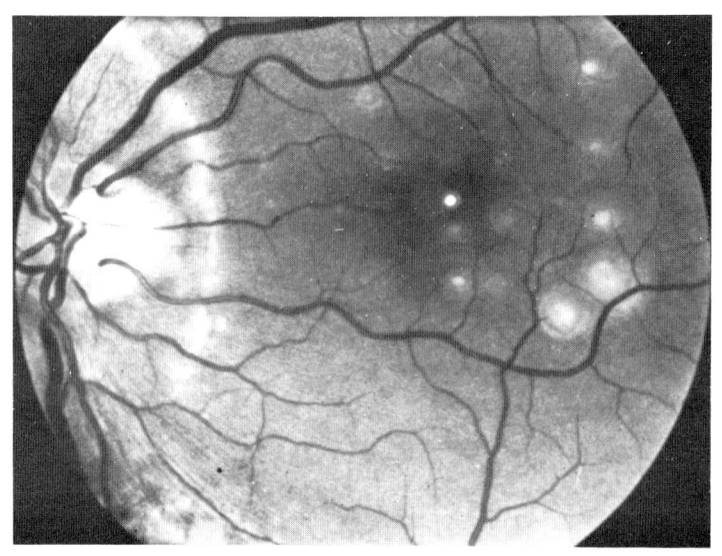

Figure 9.5. Experimental laser burns in an eye with a malignant melanoma. The burn adjacent to the optic nerve head did not involve the nerve fiber layer of the retina and thus did not reduce the patient's central visual acuity. The grid-like pattern of laser burns in the macula produced several small scotomas. The small dense laser burn in the fovea reduced the patient's visual acuity from 20/20 to 20/40.

the developing orbit. Sarcomas, particularly osteogenic and fibrosarcomas, sometimes occur in the orbit many years after irradiation to that region.

Electric Injuries

The passage of electricity of sufficient current through the ocular tissues can cause an electrothermal burn. Depending upon the duration and the amount of current, the degree of resulting coagulative necrosis varies. An inflammatory reaction develops around the necrotic tissue. An electric burn, which may be microscopic, occurs at the site where the conductor establishes contact with the body and the metallic constituents of the conductor may deposit in the tissue and discolor it. A cataract can occur in the absence of other discernible ocular abnormalities, several weeks after an electrical injury to the eye.

THERMAL INJURIES

Extremes of heat and cold are also injurious to the eye. The eye may be injured by flames, burns, or scalds. The eyelids close tightly with most

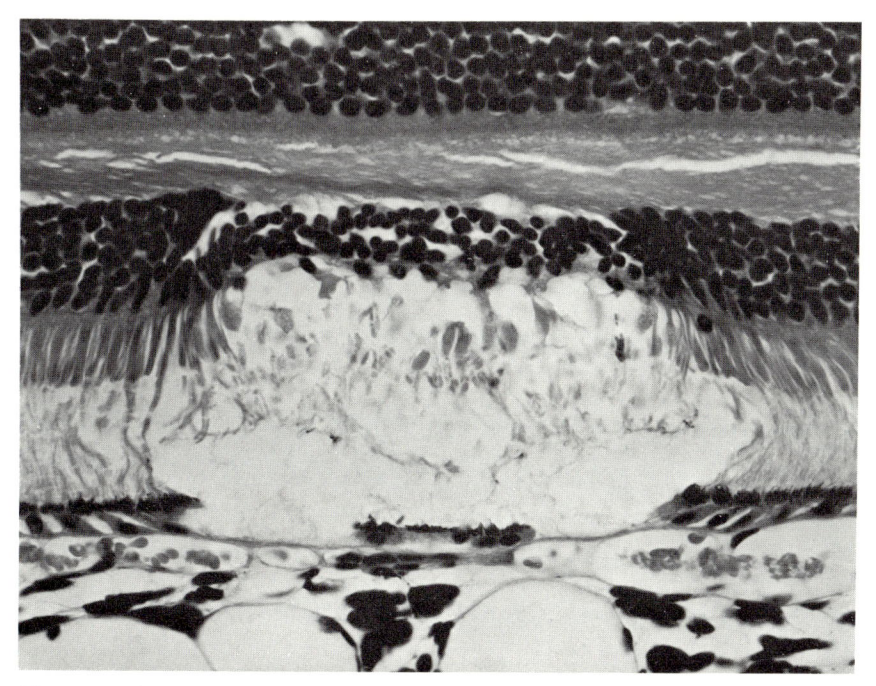

Figure 9.6. An acute chorioretinal lesion produced by the gallium arsenide laser. (H & E, × 500)

thermal burns, and as a rule, the eyelids are burnt more than the globe. Hypothermic injury can occur during cryosurgery, which is used in the treatment of cataracts and retinal detachments.

ULTRASONIC INJURIES

Sound waves of more than 20,000 cycles per minute are inaudible to man. Such ultrasonic waves are used by ophthalmologists to produce an echogram of the globe and orbit, enabling them to study the eye when the media are opaque and to examine the orbital tissues. Ultrasound has been shown to produce cataracts and other ocular lesions in experimental animals.

CHEMICAL INJURIES TO THE EYE

A wide variety of household and industrial chemicals can enter the eye and injure it. Acids tend to cause immediate damage because they become buffered by the tissue. The burns produced by alkalis are common and often result in a characteristic progressive lesion. The cornea immediately becomes opaque and in several days, new vessels begin to grow into it. After a delay of about 2 weeks, a central corneal ulcer commonly develops and often perforates. Collagenase activity has been demonstrated in alkali-burned corneas and is thought to play an important role in the genesis of the ulcer and perforation. Several chemicals cause local corrosive affects. Late sequelae of chemical burns include corneal opacification or even necrosis and perforation, as well as scarring and shrinkage of the conjunctiva.

INDIRECT OCULAR INJURIES

Physical Trauma to the Head

Aside from direct physical trauma to the eye, the ocular tissue can be affected secondary to injury elsewhere. In head injuries that result in fractures at the base of the skull, extravasated blood may seep into the orbit and appear in the subconjunctival tissue and eyelids (Fig. 9.7). Unlike hemorrhages caused by direct blows to the eye, the subconjunctival hemorrhage is most extensive posteriorly and the hemorrhage does not extend beyond the margins of the orbit. Subconjunctival and retinal hemorrhages are common in head injuries that follow a rapid deceleration of the head. These are believed to arise from the slamming of blood from the back of the head to the front.

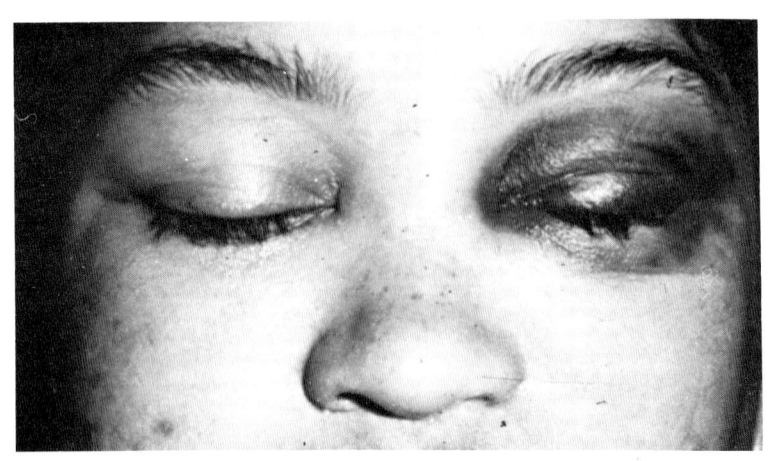

Figure 9.7. This circumscribed hemorrhage of the upper and lower left eyelids followed a fall onto the back of the head. A fracture of the base of the skull is commonly associated with such hemorrhage.

Physical Trauma to Areas Other than the Head

Hemorrhages, exudates, and ischemic axonal reactions (cotton-wool spots) sometimes develop in the retina after crushing injuries to the chest or distant parts of the body. These presumably result from several mechanisms, including some elevation of intravenous pressure, fat emboli, or retinal ischemia due to exsanguination. With strangulation, conjunctival petechiae are common.

Decompression Injuries

When persons in compressed air, such as deep sea divers, are decompressed suddenly or individuals ascend too rapidly to high altitudes, the solubilized air within their blood stream becomes released and forms bubbles of gas which embolize and sometimes occlude the circulation, including that of the eye.

REFERENCES

Brown, S. I., Weller, C. A., and Vidrich, A. The pathogenesis of ulcers of the alkali burned cornea. Arch. Ophthalmol. 83:205–208, 1970.

Burger, P. C., and Klintworth, G. K. Experimental retinal degeneration in the rabbit produced by intraocular iron. Lab. Invest. 30:9–19, 1974.

Cox, M., Schepens, C., and Freeman, H. Retinal detachment due to ocular contusion. Arch. Ophthalmol. 76:678–685, 1966.

Duke-Elder, S., and MacFaul, P. A. System of Ophthalmology, Vol 14, Part 1: Mechanical Injuries and Part 2: Non-Mechanical Injuries. Ed. by S. Duke-Elder, C. V. Mosby, Co., St. Louis, 1972.

Tso, M. O. M. Photic maculopathy in rhesus monkeys: a light and electron microscopic study. Invest. Ophthalmol. 12:17–34, 1973.

Wolbarsht, M. L. (Ed). Laser Applications in Medicine and Biology, Vol. 1. Plenum Press, New York, 1971.

10

ANOMALIES OF INTRAOCULAR PRESSURE

Normally, about 2.5 ml of aqueous humor are formed each day by the ciliary body. Fluid passes from the capillaries within the ciliary processes into the surrounding stroma and hence through the two layers of the ciliary epithelium to the posterior chamber. From the posterior chamber, the aqueous humor circulates through the pupil into the anterior chamber toward the drainage angle between the corneoscleral and uveal portions of the eye (Fig. 10.1). At this site, the aqueous drains into the spaces between the fibroelastic cores of the trabecular meshwork (spaces of Fontana). Using a special contact lens, it is possible to view the angle formed between the iris and the trabecular meshwork. Normally, this angle is between 20° and 45° and the entire trabecular meshwork is visible. Angles less than 20° are termed narrow angles and angle closure glaucoma is more probable in these eyes. The outer lining of the trabecular meshwork abuts against blind canalicular extensions of the endothelial-lined canal, which courses circumferentially about the cornea (Schlemm's canal). The two compartments are not in direct continuity with each other. The transport of aqueous humor from the trabecular meshwork to Schlemm's canal requires transport across a cellular barrier. The endothelial cells lining the trabecular wall of Schlemm's canal contain giant cytoplasmic vacuoles that are larger than those observed in almost all other mammalian cells. Each individual cell usually possesses only a single vacuole that ranges in diameter from 2–15 μ. These vacuoles probably play a role in the transport of aqueous humor from the anterior chamber to Schlemm's canal. It is possible that they create a transitory canal which may allow the passage of blood cells, other wandering cells, and particulate matter. Similar cells occur in the arachnoid villi and probably contribute to the drainage of cerebrospinal fluid. Morphologic studies of eyes, after fine tracer particles have been introduced into the anterior chamber, suggest that other ancillary outflow paths may drain aqueous humor by other routes, such as through the ciliary body and root of the iris. Thirty-odd efferent channels drain from

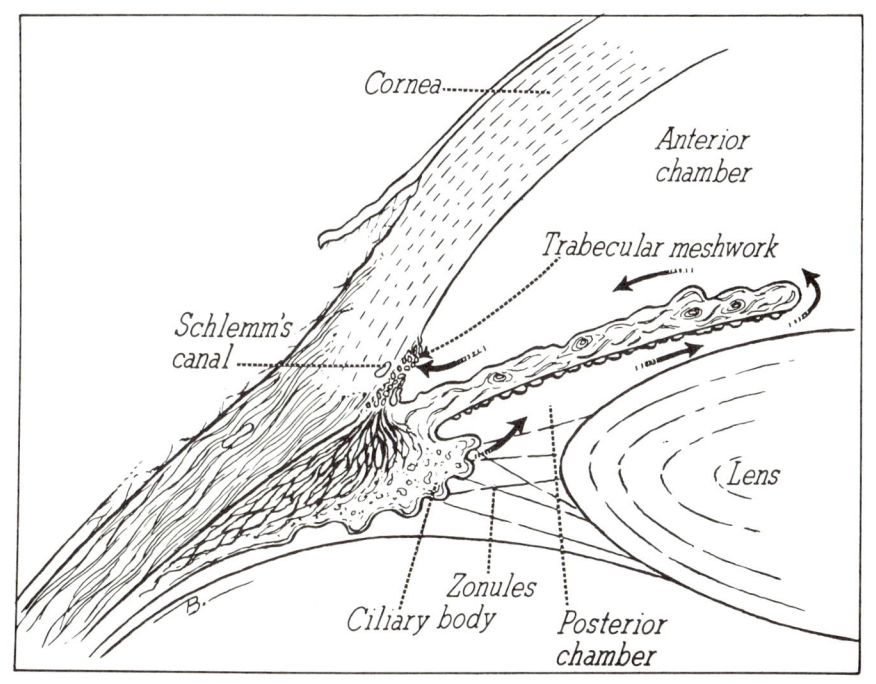

Figure 10.1 Schematic depiction of the drainage pathway of the aqueous humor.

Schlemm's canal into the aqueous veins and hence into the intrascleral veins. The aqueous is not a simple filtrate of serum due to the "blood aqueous barrier." At the site of its production, the aqueous usually contains less protein and glucose than plasma, but much more ascorbic acid. The production of aqueous humor is balanced by its drainage. There are individual variations in the rate of production and drainage of the aqueous, but the anterior portion of the adult human eye normally contains about 0.3 ml of aqueous humor within the anterior and posterior chambers.

GLAUCOMA

The intraocular pressure is produced largely by the fluid within the aqueous humor and the vitreous. It is determined in the living patient by measuring the force required to indent or flatten the cornea and it normally measures 10–20 mm Hg. Under abnormal circumstances, the pressure may be excessively high. When the intraocular pressure is sufficiently elevated to cause temporary or permanent impairment of vision or damage to the optic nerve, the term glaucoma is applied. This usually does not occur until the intraocular pressure is above 20 mm Hg.

Glaucoma most often results from an impaired outflow of the aqueous humor due to a congenital or acquired anomaly of the anterior segment of the eye. The mechanical obstruction may occur between the iris and lens, in the angle of the anterior chamber, in the trabecular meshwork, in Schlemm's canal, or in the venous drainage of the eye. In theory, glaucoma might also result from the hypersecretion of aqueous humor with normal aqueous outflow paths. However, this type of glaucoma is extremely rare.

Glaucoma has traditionally been divided into several varieties. If it occurs in the absence of obvious ocular disease, it is designated primary glaucoma. In this type, the anterior filtration angle may be patent (open angle glaucoma) or closed (closed angle glaucoma). When glaucoma complicates other antecedent or concomitant ocular disease, it is termed secondary glaucoma. This may follow traumatic, vascular, inflammatory, neoplastic, or developmental anomalies. In contrast to primary glaucoma, only one eye is commonly affected in secondary glaucoma.

Histopathology of Glaucoma and Effects of Ocular Hypertension on Tissue

Eyes of patients with glaucoma show abnormalities related to the underlying ocular disease, to alterations due to increased intraocular pressure, and sometimes, also the effects of therapeutic procedures that have been performed to alleviate the glaucoma. The effect of increased intraocular pressure varies with the age of the patient when the intraocular pressure begins to rise and with the rate at which the pressure becomes elevated. There are also individual variations in the tolerance of the eyes to the adverse effects of high intraocular pressure. Some individuals seem to be able to tolerate an intraocular pressure at least three standard deviations above the mean (24 mm Hg) for many years, without developing a visual field loss or optic atrophy. Individual variations probably depend to some extent upon the adequacy of the circulation of the retina and optic nerve head. Elevated intraocular pressure results in secondary effects on the ocular tissues. Edema of the corneal epithelium and stroma occur especially with an acute elevation in intraocular pressure. The corneal edema, which reflects dysfunction of the corneal endothelial cells and sometimes tears in Descemet's membrane, results in opacification of the cornea and sometimes subepithelial bullae (bullous keratopathy). The epithelium may degenerate and desquamate. With chronic glaucoma the entire uveal tract, including the ciliary and iridial muscles, becomes atrophic. Also, connective tissue may invade between Bowman's zone and the the corneal epithelium (glaucomatous pannus), and ectropion uveae sometimes develops. Degenerative changes in the iris liberate melanin granules into the aqueous. The pigment deposits in sites influenced by convection currents in the aqueous. It collects in the filtration angles of the anterior chamber and in the corneal endothelial cells, as well as on the surface of the iris, lens, and zonules. It may deposit on the posterior surface of the cornea as a spindle-shaped vertical line (Krukenberg's spindle). Numerous discrete subcapsular opacities appear in the pupillary region of the lens ("glaukomflecken"), after an acute rise in intraocular pressure. These flecks correspond to areas of focal necrosis of the subcapsular epithelium. Chronic glaucoma in the adult results in no apparent pathologic change in the sclera or cornea for years.

An acute rise in intraocular pressure results in a congested edematous uveal tract. Acute angle closure glaucoma is accompanied by congestion of the deep pericorneal vascular plexus. In infants, the corneoscleral coat is usually plastic and intraocular hypertension stretches and thins the outer ocular coat. The whole eye commonly enlarges, giving rise to a total scleral staphyloma and buphthalmos. Cupping of the optic disc is often less than that which occurs in adults with prolonged glaucoma. Prolonged intraocular hypertension causes a loss of nerve fibers and an ectasia, usually with a prominent overhanging lip at the optic disc. This leads to a characteristic cupped excavation of the optic disc (glaucomatous cupping) (Figs. 10.2, 10.3), accompanied by a nasal displacement of the retinal blood vessel. Ischemic changes produced by an elevated intraocular pressure relative to intravascular capillary pressure are thought to damage the optic disc. Edema of the optic disc may follow an episode of acute glaucoma that lasts several days.

Increased intraocular pressure causes degenerative changes in the retina. These characteristically become evident first in the ganglion cell and nerve fiber layers. Although the outer retina remains relatively normal initially, it also eventually degenerates in long standing glaucoma. As a rule, the corneoscleral envelope is resistant to stretching in individuals over the age of 3 years and there is rarely more than moderate thinning. A sustained ocular hypertension leads to a prominent bulging of the cornea or sclera at weak points, such as sites of scars in the outer coat of the eye. Scleral staphylomata occur mainly at two sites: covering the ciliary body (ciliary staphyloma) and anterior to the ciliary body when extensive peripheral synechiae are present (intercalary staphyloma) (Fig. 10.4). These localized thinnings are lined by atrophic uveal tissue. Rarely do staphylomata develop near the equator of the globe where the vortex veins perforate the sclera (equatorial staphyloma).

184   *The Eye*

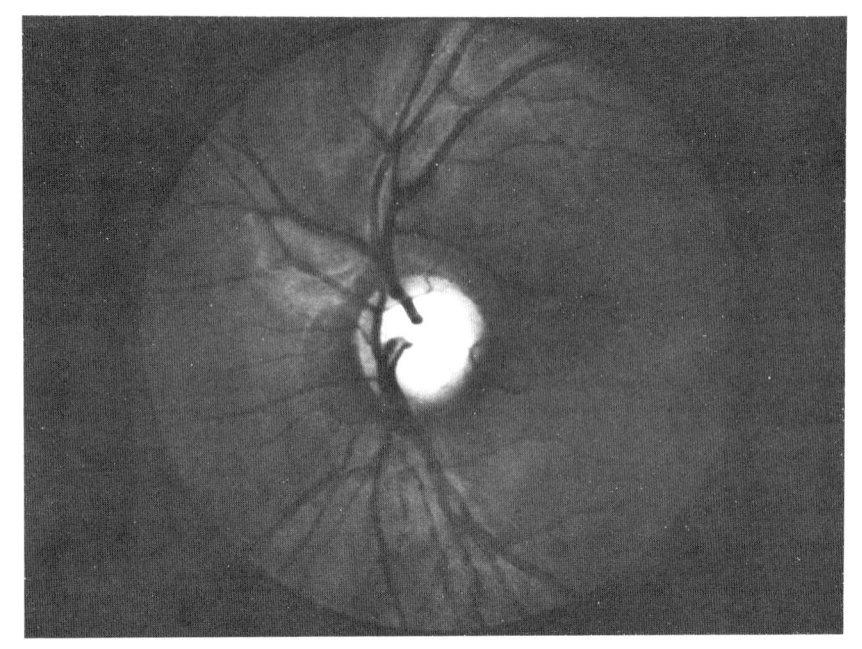

Figure 10.2 Glaucomatous cupping with nasal displacement of the blood vessels.

In chronic glaucoma, optic atrophy with a loss of axons, gliosis, and thickening of the pial septa may follow the retinal degeneration and damage to the nerve fibers at the optic disc. With an acute episode of intraocular hypertension, a characteristic variety of optic atrophy, with cavernous spaces containing hyaluronic acid, sometimes occurs (cavernous optic atrophy, Schnabel's optic atrophy).

Special Types of Glaucoma

Congenital Glaucoma (Buphthalmos)

The term congenital glaucoma has traditionally referred to glaucoma due to an obstruction of the aqueous drainage by developmental anomalies, even though the ocular hypertension may not be elevated until sometime after birth. Glaucoma due to developmental anomalies usually becomes manifest at birth or during early infancy, but occasionally the intraocular pressure may not become elevated until after 6 years of age. When glaucoma results from a developmental anomaly, it is often characterized by large eyes with steamy corneas and photophobia (Fig. 10.5). Congenital glaucoma usually results from a developmental anomaly of the angle of the anterior chamber.

Usually congenital glaucoma involves both eyes and most cases are

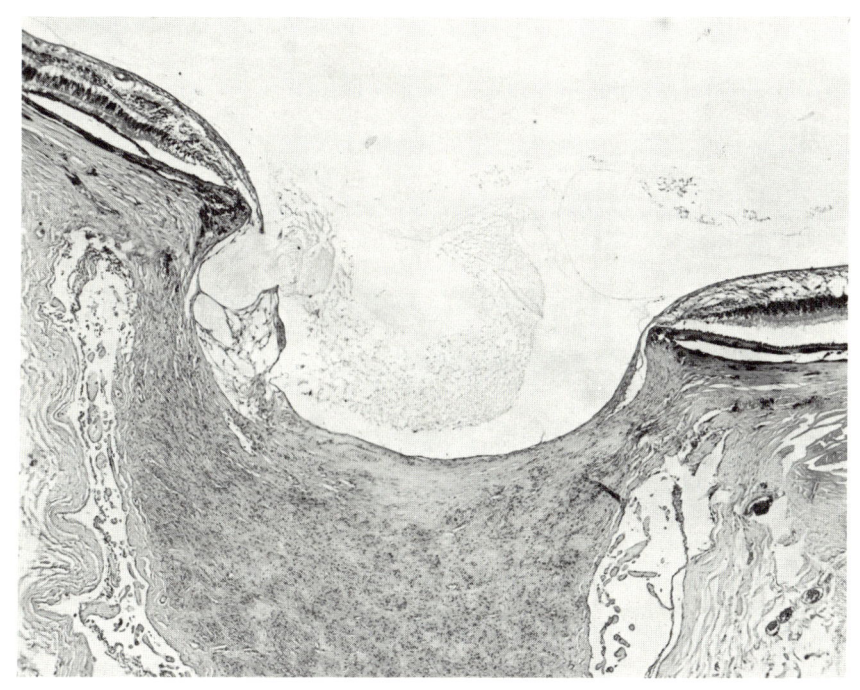

Figure 10.3. This histologic section through the optic disc shows marked cupping due to long standing glaucoma. (H & E, × 25)

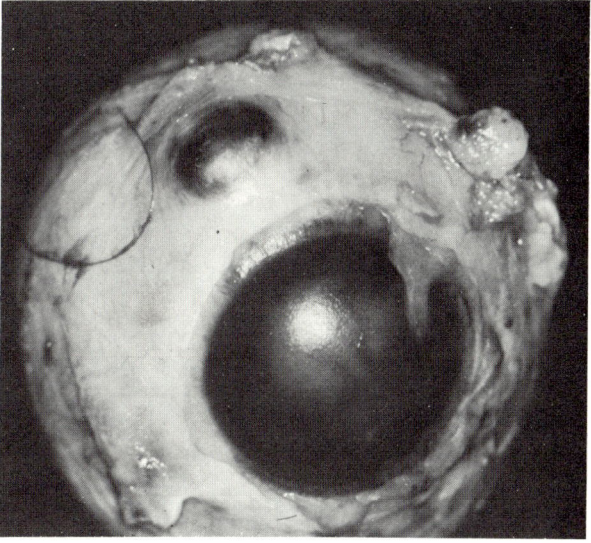

Figure 10.4. This eye was enucleated because of advanced glaucoma (absolute glaucoma) that failed to respond to treatment. Note the localized dark bulge (ciliary staphyloma) which followed the prolonged increase in intraocular pressure.

186   *The Eye*

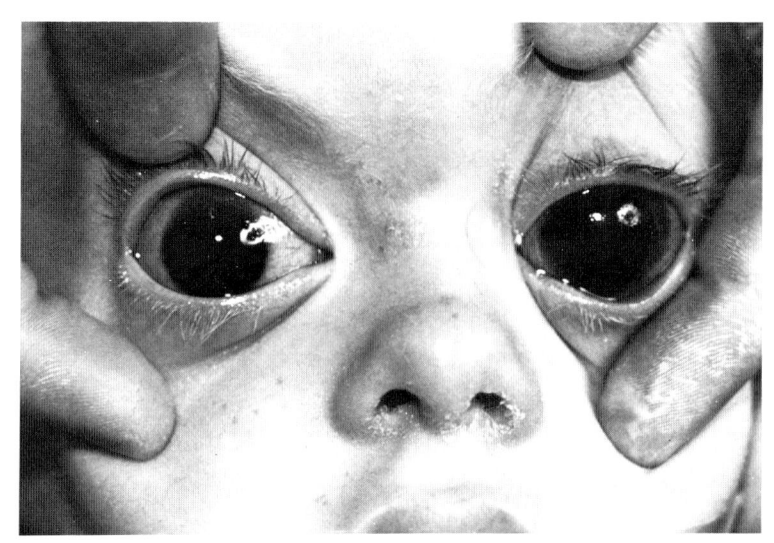

Figure 10.5. Buphthalmos secondary to congenital glaucoma. The right eye still has useful vision. The left eye is blind with a hazy cornea.

males (60–70%). The developmental anomaly that results in glaucoma may be limited to the angle of the anterior chamber. This type of congenital glaucoma is associated with a deep anterior chamber and is usually inherited as an autosomal recessive condition. Congenital glaucoma may occur as an isolated ocular anomaly. It is also often associated with a variety of malformations, especially of the iris (aniridia, hypoplasia, coloboma), cornea (microcornea, Peter's malformation, cornea plana), or lens (spherophakia, ectopia lentis). The glaucoma may be a component of several syndromes, viz. congenital rubella, Sturge-Weber, Lowe's oculocerebrorenal, Pierre-Robin, Marchesani, homocystinuria, Marfan, Axenfeld, Rieger's anomaly, oculodentodigital dysplasia, trisomy 13, and trisomy 18. The Sturge-Weber syndrome is the most common cause of monocular glaucoma in infants in the United States and several other countries.

Primary Angle Closure Glaucoma

A type of glaucoma designated primary angle closure glaucoma (closed angle glaucoma, narrow angle glaucoma, acute glaucoma) occurs especially after the age of 40 years in individuals who possess an abnormally narrow angle in which the peripheral iris is very close to the inner surface of the trabecular meshwork. The peripheral iris commonly inserts at the extreme anterior edge of the ciliary body. The cornea is often small. The condition occurs especially in hyperopes with a large ciliary body and a smaller than normal eye. The anterior chamber may become further narrowed as the lens normally thickens with age. A familial tendency to narrow, shallow angles is frequently present. The disorder is characterized clinically by sudden episodes of intraocular hypertension during which pain occurs and halos or rings are seen around lights. These attacks occur when mechanical obstruction to the outflow of the aqueous humor results in elevations in intraocular pressure. Elevated intraocular pressure may be precipitated by pupillary dilation, when the filtration angle becomes occluded by the apposition of the peripheral iris to the trabecular meshwork. Elevated intraocular pressure also may occur if the pupils become blocked, for example by a swollen lens, and aqueous accumulates in the posterior chamber, ballooning the peripheral iris forward (iris bombé). The intraocular pressure is normal between attacks. The condition is bilateral, although it may become apparent in one eye 2-5 years before its fellow. After numerous episodes, especially if prolonged, adhesions ultimately develop between the iris and the trabecular meshwork and cornea (peripheral anterior synechiae). When this occurs, it accentuates the blockage of the outflow of the aqueous humor.

Primary Open Angle Glaucoma (Chronic Simple Glaucoma)

This variety of glaucoma occurs principally in the 6th decade and is estimated to be present in 1-3% of the population over the age of 40. There is no sex predisposition. It is a major cause of blindness in the adult in the United States. There is an insidious elevation of intraocular pressure which is initially asymptomatic. The angle is open gonioscopically. There is apparently an increased resistance to the outflow of the aqueous humor in the chamber angle. It is generally believed that the site of maximal resistance is in the area adjacent to Schlemm's canal. Although almost always bilateral, one eye may be affected more severely than the other. Pressures are not as high as in narrow angle glaucoma. Tissue sections have been studied in relatively few eyes in the early stages of the condition and the basic lesion is still unclear.

Secondary Glaucoma

Glaucoma may follow trauma because of one of several mechanisms. These include direct injury to the trabecular meshwork, intraocular hemorrhage, complications of wound healing or inflammation (e.g. peripheral anterior synechiae), and obstruction of the trabecular meshwork by displaced lens material. After a corneal graft or cataract extraction, the delayed reformation of the anterior chamber may result in peripheral anterior synechiae and glaucoma. Nonperforating injuries may lead to increased intraocular pressure because of vascular congestion. Monocular open angle glaucoma may occur 10–20 years after an ocular contusion which causes recession of the angle and ciliary body with damage to the trabecular meshwork.

A hypermature cataract leaks lens material into the aqueous humor and may result in the sudden onset of glaucoma ("phacolytic glaucoma"). Cellular or necrotic debris or extruded lens material, which may be present within macrophages, can plug up the trabecular meshwork.

The drainage of aqueous humor through the chamber angle may become obstructed by several mechanisms: the corneal or conjunctival epithelium may enter the anterior chamber after intraocular surgery or aspirations, or after injury to the cornea or corneoscleral limbus ("epithelial downgrowth"); fibrous tissue may extend through a wound in the inner aspect of the cornea ("fibrous downgrowth") and corneal endothelial cells may grow across the angle and form abnormally located Descemet's membrane.

The pupil may become obstructed by adhesions between the iris and lens, as after iritis, as well as by a spherical lens (pupillary block glaucoma). If this prevents the flow of aqueous from the posterior to the anterior chamber, the iris bulges forward (iris bombé) and may even come into apposition with the angle. Angle closure glaucoma also can result from pressing of the iris and ciliary body forward, as with swollen cataractous lens ("intumescent cataract") (phacomorphic glaucoma). Displaced lenses are frequently associated with glaucoma. In Marfan's syndrome, in which the lens is dislocated posteriorly, a developmental anomaly of the anterior chamber is common. A common cause of glaucoma in traumatic posterior lens dislocation is the damage that occurs to the trabecular meshwork at the time of the ocular injury.

A rise in intraocular pressure may follow a retarded outflow of aqueous due to an increased episcleral venous pressure. This can occur with an obstruction to the venous drainage of the eye or with a carotid-cavernous fistula. Interference with the venous drainage of the eye can occur with orbital and mediastinal masses. When venous obstruction is released, the intraocular pressure returns to normal.

The filtration angle of the anterior chamber, including the trabecular meshwork, may become occluded by peripheral anterior synechiae, as after iritis or rubeosis iridis; granulomas of the iris or ciliary body; particulate matter, like cellular or necrotic debris which may be present within macrophages; massive hyphema; or by an anterior dislocated lens. Neoplasms, such as melanomas, can cause glaucoma by several different mechanisms. These include neoplastic invasion of the trabecular meshwork, rubeosis iridis, and obstruction of the filtration angle by intra- and extracellular necrotic debris. When numerous melanin granules are evident in the angle, the term pigmentary glaucoma is used.

Glaucoma may accompany therapy with α-chymotrypsin or steroids.

"Low Tension Glaucoma"

For a considerable time, ophthalmologists have recognized individuals with the characteristic visual field defect and all of the ophthalmoscopic features of chronic simple angle glaucoma, including optic nerve head excavation and atrophy, but without elevation in intraocular pressure. This entity, called low tension glaucoma (soft glaucoma, pseudoglaucoma), is fairly common in elderly individuals and is due almost certainly to infarction of the optic nerve head. Of the many causes, atherosclerosis and giant cell arteritis are the most frequent. It is probable that glaucomatous damage to the optic nerve is based on a relative vascular insufficiency of optic nerve head. The changes in the visual field have been noted to occur in hypertensive patients when their blood pressure has been rapidly lowered by antihypertensive therapy. It has also been suggested that some eyes might be hypersensitive to a normal intraocular pressure.

HYPOTONY (HYPOTONIA)

Decreased intraocular pressure may follow accidental penetrating wounds, intraocular surgery, and uveitis. Edema of the optic disc, retina, ciliary body, and choroid may result presumably because of an impaired absorption of fluid due to low intraocular pressure. Folds frequently develop in Descemet's membrane due to corneal edema, causing backward swelling of cornea and the sensory retina may become detached from the retinal pigment epithelium. Prolonged ocular hypotension may herald atrophy of the globe.

REFERENCES

ANDERSON, D. R. Pathology of the glaucomas. Br. J. Ophthalmol. 56:146-156, 1972.
ARMALY, M. F., BECKER, B., HAAS, J. S., KRONFELD, P. C., POLLACK, I. P., SHAFFER, R. N., AND ZIMMERMAN, L. E. Symposium on Glaucoma Transactions of the New Orleans Academy of Ophthalmology C. V. Mosby, Co., St. Louis, 1967.
DUKE-ELDER, S. (Ed). System of Ophthalmology. Diseases of the Lens and Vitreous: Glaucoma and Hypotony, Vol. II. C. V. Mosby, Co., St. Louis, 1969.
HAYREH, S. S. Blood supply of the optic nerve head. Br. J. Ophthalmol. 53:721-748, 1969.
KOLKER, A. E., AND HETHERINGTON, J., JR. (Eds). Becker-Shaffer's Diagnosis and Therapy

of the Glaucomas, 3rd edition. C. V. Mosby, Co., St. Louis, 1970.

LAMPERT, P. W., VOGEL, M. H., AND ZIMMERMAN, L. E. Pathology of the optic nerve in experimental acute glaucoma: electron microscopic studies. Invest. Ophthalmol. 7:199-213, 1968.

SHAFFER, R. N., AND WEISS, D. I. Congenital and Pediatric Glaucomas, C. V. Mosby, Co., St. Louis, 1970.

TRIPATHI, B. J., AND TRIPATHI, R. C. Vacuolar transcellular channels as a drainage pathway for cerebrospinal fluid. J. Physiol. 239:195-206, 1974.

YANOFF, M. Glaucoma mechanisms in ocular malignant melanomas. Am. J. Ophthalmol. 70:898-904, 1970.

11

MISCELLANEOUS CONDITIONS OF THE EYE

CONJUNCTIVA AND CORNEA

The conjunctiva, one of the most exposed mucous membranes, contains the caruncle in its inner angle. The caruncle is covered by a nonkeratinized, stratified squamous epithelium with hair follicles, large sebaceous glands, and colorless hairs. It is prone to staphylococcal infection. Subepithelial lymphoid tissue is present in the bulbar conjunctiva, the orbital part of the palpebral conjunctiva, and the plica semilunaris.

The cornea is the transparent portion of the outer layer of the globe. Its epithelium does not normally keratinize and is covered by a tear film. The corneal stroma contains fibroblasts, as well as lamellae of collagen fibers, chondroitin-4-sulfate, and keratosulfate. The normal cornea possesses no blood vessels or lymphatics.

Opacification of the crystal clear cornea follows the accumulation of fluid within it. A characteristic of the cornea in the living subject is its ability to maintain a deturgescent state, despite a marked hydrophilic property of the corneal stroma. The corneal endothelium plays a crucial role in this regard and corneal edema complicates numerous disorders that affect the corneal endothelium. These include increased intraocular pressure, donor tissue for corneal grafts which either lacks or has an abnormal corneal endothelium, and an idiopathic disease of the corneal endothelium (*Fuchs' epithelial-endothelial corneal dystrophy*). The latter condition occurs mainly in middle-aged to elderly women. Due to the failure of the endothelium to remove water from the corneal stroma, there is epithelial edema especially in the central cornea. There is a thickened Descemet's membrane with excrescences (*cornea guttata*) (Fig. 11.1). Whenever the corneal endothelium becomes damaged, vesicles develop especially beneath the corneal epithelium (bullous keratopathy) (Fig. 11.2). Swelling of the stroma also occurs and may lead to marked fold in Descemet's membrane.

192   The Eye

Figure 11.1. Numerous excrescences (cornea guttata) are evident in this posterior view of the cornea from a patient with Fuchs' epithelial-endothelial corneal dystrophy. (× 62.5)

Keratoconus

The condition of keratoconus (conical cornea) is characterized by an ectasia of the axial portion of the cornea (Fig. 11.3). It is usually bilateral and has an onset in youth or adolescence. The cause and pathogenesis are obscure. Dehiscences in the basement membrane of the corneal epithelium and Bowman's zone are common and corneal edema frequently develops.

Pinguecula

A slightly elevated, yellowish gray, oval-shaped area commonly develops in the conjunctiva near the limbus in the interpalpebral fissure. It may be on either side of the globe, but is more frequently situated nasally (Fig. 11.4). Because the abnormality was at one time thought to be composed of fat, it has become known as a pinguecula (Greek: *pinguecula*, fat). It is usually bilateral and increases in incidence with age. The lesion consists of an elastotic degeneration of the connective tissue. It is not digested with elastase. The location of the disorder, together with its morphologic similarity to actinic elastosis, suggests that it is also due to chronic exposure to actinic rays from the sun.

Pterygium

A pterygium consists of a triangular fold of vascularized bulbar conjunctiva which encroaches in the shape of an insect wing in the interpalpebral fissure onto the cornea (Greek: *pterygium*, wing) (Fig. 11.5). It may occur on either side of the cornea but is usually located nasally. If it extends across the pupil, vision may be impaired. It is often associated with a pinguecula. Pterygia are common in the tropics and subtropics, particularly among individuals exposed to chronic irritation by dust, wind, and sunlight. The subepithelial connective tissue invades the superficial cornea and destroys Bowman's layer. After excision, recurrences are common, particularly in tropical areas.

Disorders of the Skin

The skin, conjunctiva, and cornea not only share a superficial location and similar developmental origin, but may be involved in a variety of

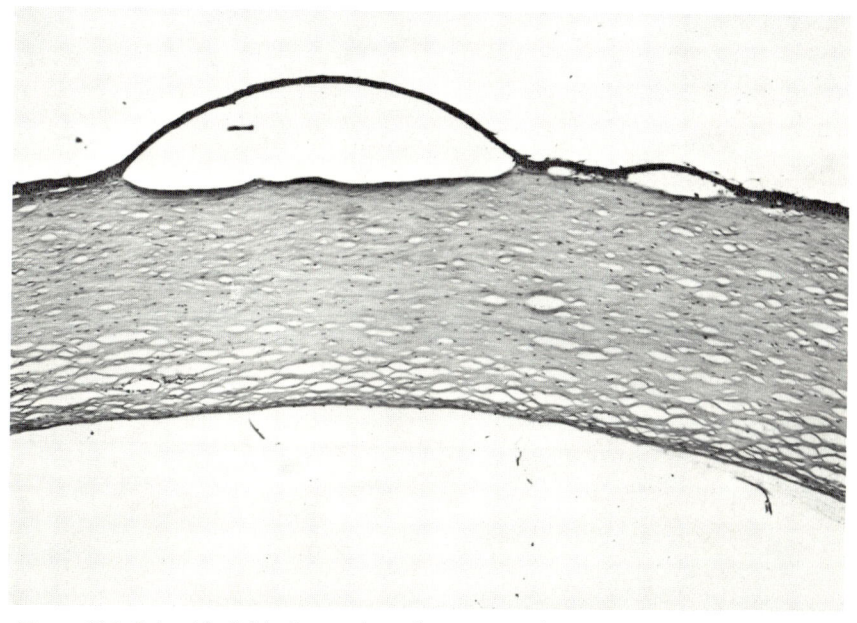

Figure 11.2. Subepithelial bullae such as these commonly accompany corneal disorders in which the corneal endothelium is abnormal. (H & E, × 37)

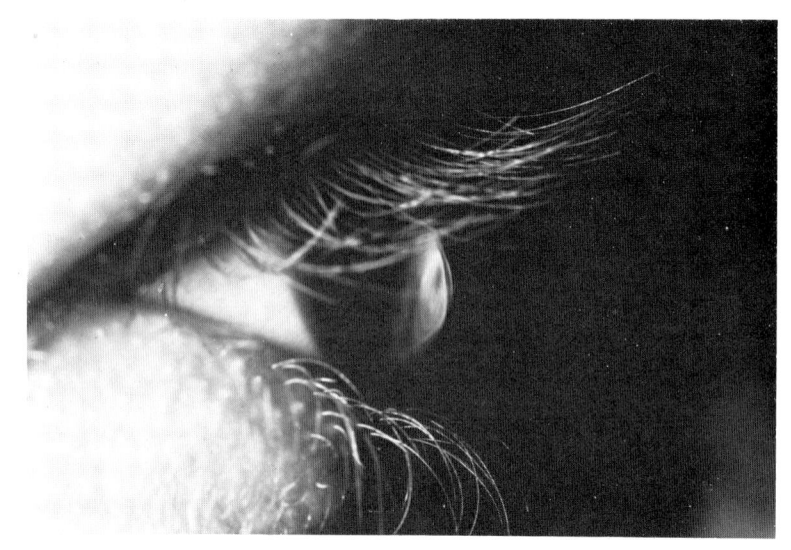

Figure 11.3. Keratoconus. When viewed from the side, the central cornea can be seen to bulge forward.

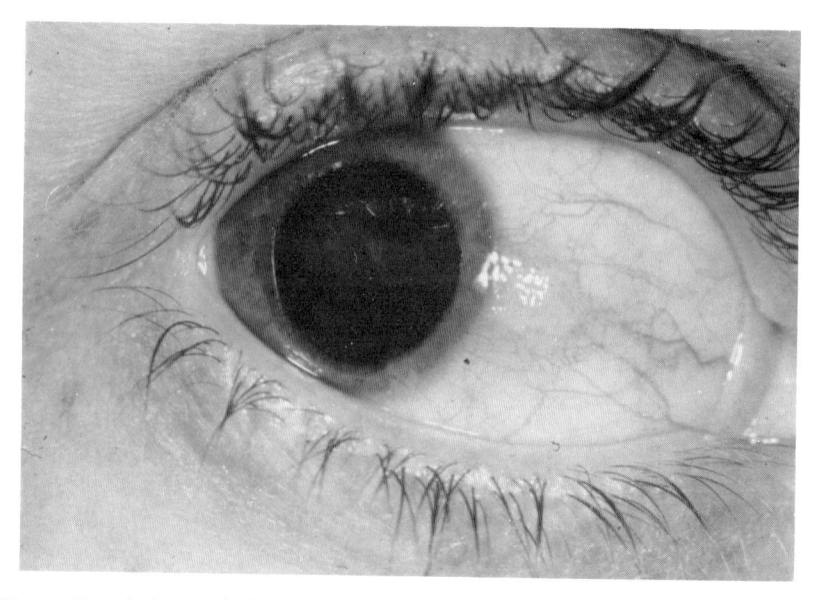

Figure 11.4. A pinguecula located at the nasal limbus. These small elevated lesions are often bilateral.

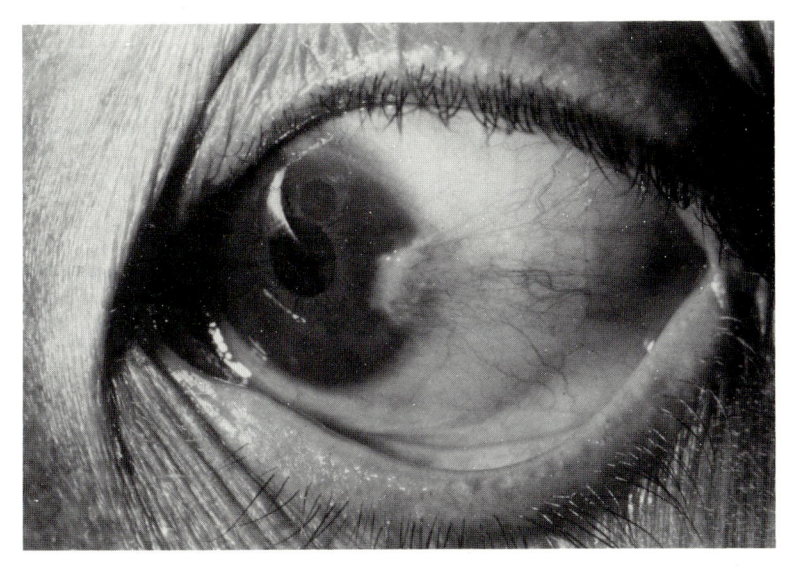

Figure 11.5. A pterygium at the temporal limbus. The cornea is often abnormal immediately in front of the advancing edge of vascularized tissue.

similar disorders. These include *rosacea, pemphigus, erythema multiforme (Stevens-Johnson syndrome), dermatitis herpetiformis, epidermolysis bullosa, erythema nodosum, psoriasis, atopic dermatitis,* and *xeroderma pigmentosum.*

Pemphigus Vulgaris and Benign Mucous Membrane Pemphigoid

Two different diseases which frequently cause confusion are pemphigus vulgaris and benign mucous membrane pemphigoid. Pemphigus vulgaris, a blistering cutaneous disease of adults characterized by recurrent bullae, may involve the oral and even the conjuctival mucosa. The bullae are predominantly subepithelial and usually heal without scarring, unless the lesion becomes secondarily infected. The condition is usually fatal in the absence of treatment. Benign mucous membrane pemphigoid, on the other hand, involves the conjunctiva and other mucous membranes, but seldom affects the skin. It appears insidiously after the age of 60 in recurrent episodes. This chronic disease usually affects both eyes. Unlike pemphigus vulgaris, scarring and shrinkage of the conjunctiva occur in benign mucous membrane pemphigoid. The cicatrization sometimes occurs to the extent that the eyelids become firmly adherent to the globe. The conjuctival secretion may diminish because of destruction of conjunctival goblet cells. The cornea often becomes opaque and vascularized. Although the patient's general health is not affected, the conjunctival scarring often blinds patients who live long enough.

Neuropathic Keratopathy

A keratopathy, characterized by a diffuse corneal edema, epithelial bullae, and indolent ulceration which may perforate, can follow anesthesia of the cornea as with lesions of the trigeminal nerve. The condition is not directly due to the corneal anesthesia because it does not always follow Gasserian ganglion section and can be prevented by suturing the eyelids together. The insensitivity of the cornea results in a loss of the reflex protective mechanism of the cornea and predisposes it to trauma and desiccation.

Keratoconjunctivitis Sicca

The secretions of sebaceous glands (Meibomian glands, Zeis' glands), sweat glands (Moll's glands), lacrimal and accessory glands, and the goblet cells of the conjunctiva lubricate the eye. A defective formation or excessive evaporation of these secretions exposes the cornea to desiccation and results in a characteristic corneal and conjunctival reaction (*keratoconjunctivitis sicca*). Fissures develop in the dry corneal and conjunctival epithelium which lacks its normal luster. The damaged cells, which desquamate and form punctate erosions, are readily demonstrated in the living patient by staining with fluorescein or rose bengal. The corneal epithelium may desquamate as filamentous threads (*filamentous keratopathy*). Secondary infection often ensues.

Exposure Keratopathy

The conjunctival and corneal epithelium may become keratinized after prolonged failure of the lids to cover the eye. The exposed dry cornea and conjunctiva are vulnerable to dust and desiccation.

UVEA

The uveal tract comprises the pigmented and predominantly vascular portions of the eye. It includes the choroid and parts of the iris and ciliary body. The iris is lined posteriorly by a double layer of pigmented epithelial cells which contain densely packed melanin granules. The iris contains smooth muscles which are of neural ectodermal origin: the

Miscellaneous Conditions 197

sphincter pupillae located near the pupillary margin and the dilator pupillae which lies anterior to the pigment epithelium. The ciliary body consists of about 70 sagittally oriented folds (pars plicata) and a nonplicated portion (pars plana), which meets the retina in a scalloped manner at the ora serrata (Fig. 11.6). The ciliary processes contain the vascular stroma and are the main site of formation of the aqueous humor. The ciliary body contains several bands of smooth muscle. It has a double-layered surface epithelium, the outer of which is heavily pigmented, the inner slightly pigmented. Each epithelial layer rests on its own basement membrane. In the ciliary processes, apical zonular occludentes junctions form continuous or overlapping belts between the adjacent nonpigmented epithelial cells. No such junctions are present between the pigmented epithelial cells, and intravascular markers, such as horseradish peroxidase, can penetrate between the pigmented epithelial layers, but not get through the nonpigmented layers. The choroid is that portion of the uvea posterior to the ora serrata. It is highly vascular and contains nerves, melanophores, and occasional mast cells.

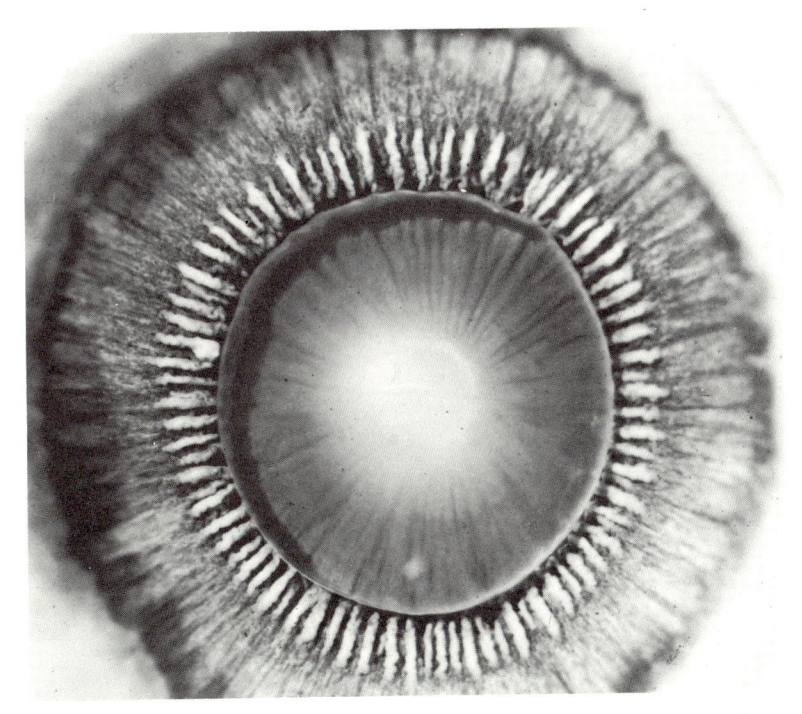

Figure 11.6. The sagittally orientated folds of the ciliary body are shown between the lens and that portion of the ciliary body that is not plicated (pars plana). ($\times$ 4.1)

198   The Eye

The fundi of black races darken with maturity because of increasing pigmentation of the choroid.

Angioid Streaks

On funduscopy, breaks in Bruch's membrane may result in a bizarre network of lines which involve especially the posterior pole (Fig. 11.7). They are initially red, but later become darker and often black. Like the retinal blood vessels, they taper peripherally, but lack dichotomous branches. They are frequently associated with hemorrhage and focal chorioretinal atrophy. Granulation tissue often grows through the defect in Bruch's membrane. When located beneath the macula, a "disciform degeneration of the macula" and loss of central vision may result. There is an increased incidence of the condition in pseudoxanthoma elasticum, Paget's disease of bone, Ehlers-Danlos syndrome, and sickle cell disease.

Ciliochoroidal Detachment

The choroid and ciliary body may separate from the sclera due to an accumulation of a transudate, exudate, or hemorrhage. A choroidal detachment also may occur secondary to traction as in phthisic eyes. The condition may follow any intraocular surgery or low intraocular tension. When this occurs, the choroid is detached, usually anterior to the equator of the eye. It is thought to result from a transudate in the suprachoroidal space. The choroidal detachment generally subsides spontaneously.

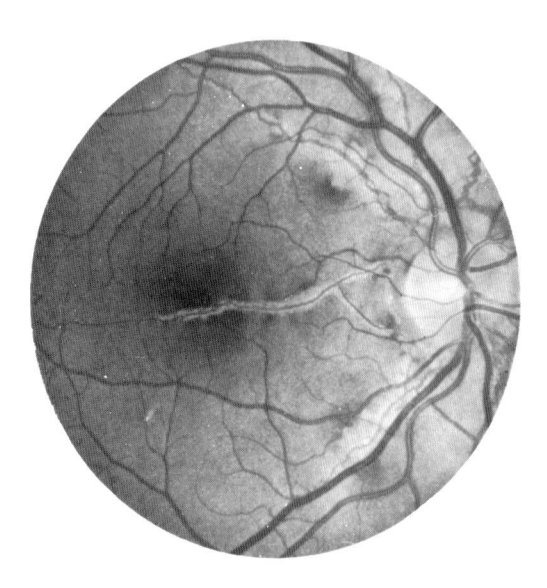

Figure 11.7. Angioid streaks. Dark lines closely resembling blood vessels radiate outward from the optic nerve. These breaks in Bruch's membrane often extend through the macula. Macular hemorrhages with loss of central vision often occur.

Heterochromia Iridis

The color of the iris depends on the quantity of pigment in its stroma and not on the pigment epithelium. Blonds have very little pigment in the iridial stroma, whereas the black races have much pigment in this location. An adrenergic innervation of the iris is essential for the development and maintenance of melanin pigment in the iris. A quantitative difference in the amount of pigment within the iris of each eye may cause each iris to be of a different color (heterochromia iridis). This may result from inflammation, an abnormality in the sympathetic innervation, or an iron-containing foreign body within the eye. Heterochromia may be associated with Romberg's syndrome of hemifacial atrophy and Horner's syndrome. In Waardenburg-Klein's syndrome (wide spaced medial canthi, flat nasal root, confluent eyebrows, white forelock, heterochromia iridis), two colors may occur within the same iris (Fig. 11.8).

Essential Atrophy of the Iris

An idiopathic, unilateral, gradually progressive atrophy of the iris affects females predominantly. Holes appear in the iris stroma which becomes progressively rarefied. The iris becomes distorted and drawn to one side, leading to an eccentric pupil. Glaucoma may eventually develop in the affected eye in this rare condition.

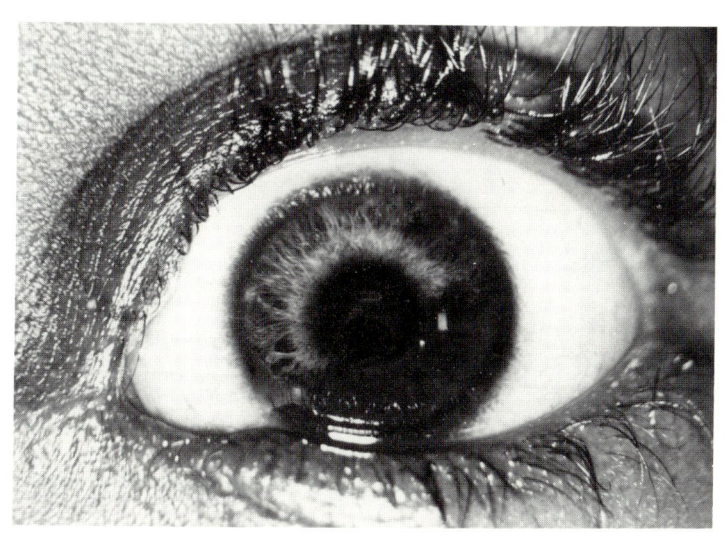

Figure 11.8. Waardenburg-Klein's syndrome. This patient has a dual colored iris in both eyes. The superior portion of the iris is blue and the inferior portion brown.

200   *The Eye*

RETINA

The retina contains the sensory nerve endings that respond to light, numerous, tightly packed neurons of variable type, neuroglia (Müller's cells), and blood vessels. Like the cerebral cortex, it contains no appreciable extracellular space. It can be divided into a number of layers (Fig. 11.9). The photoreceptors are of two types: rods which contain abundant photoreceptor pigment that is sensitive to low levels of illumination as that which occurs at night and cones which contain less pigment than rods, allow better visual acuity in the day, and permit color perception. It is estimated that there are about 120 million rods and 5

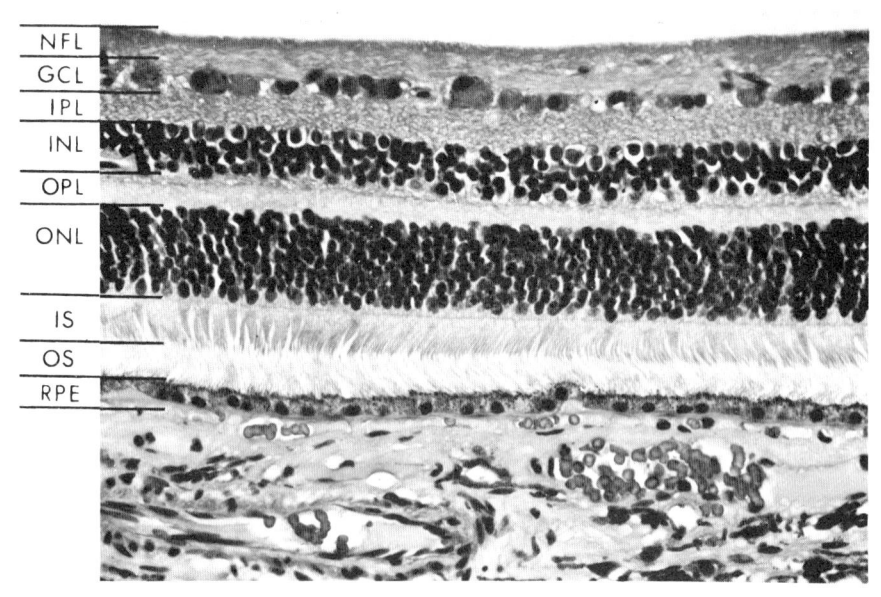

Figure 11.9. The cells and their processes are organized in several well defined layers within the retina. The nerve fiber layer (*NFL*) contains axons of ganglion cells, blood vessels, neuroglia, and Müller's fibers; the ganglion cell layer (*GCL*) has large neurons; the inner plexiform layer (*IPL*) contains the synapses between the bipolar and ganglion cells, as well as a few glial cells and blood vessels; the inner nuclear layer (bipolar layer) (*INL*) contains nuclei of the bipolar cells, the nuclei of Müller's cells, the association cells (the large monopolar amacrine and horizontal cells), and a few ganglion cells; the outer plexiform layer (*OPL*) contains synapses between the bipolar layer and the photoreceptors; the outer nuclear layer (*ONL*) contains nuclei and cell bodies of the rods and cones. The cone nuclei are in the outermost part of the layer. The cell bodies are separated by the radial glia of the retina (Müller's cells). The sensory nerve endings are divided into inner segments (*IS*) and outer segments (*OS*). The sensory retina is closely applied to the retinal pigment epithelium (*RPE*), but can be readily separated from it. (H & E, × 250)

million cones in man. The rods and cones can be divided into an outer and inner segment. The disc-like appearance of the outer segment of cones is caused by numerous invaginations of the cell membrane. The outer segment of rods contain a stack of double-layered membranous discs which contain rhodopsin (visual purple). The inner segments contain numerous mitochondria and are connected to the outer segment by cilia which contain nine pairs of tubules arranged in a ring. The rod outer segment discs are continually renewed and the pigment epithelium is responsible for removing the terminal discs. In the monkey eye, each retinal rod produces 80–90 discs per day and the entire complement of outer disc segments is replaced every 1–2 weeks. It has been calculated that each retinal pigment epithelial cell engulfs and destroys about 3000 discs every day. The neurosensory retina can be divided into several parts. The retina terminates at the ora serrata from which site the sensory retina continues forward as the "nonpigmented" ciliary epithelium. The Müller cells invest the cell bodies and processes of the neurons, as well as retinal blood vessels. Apical projections of the Müller cells extend beyond the external limiting membrane. There is a depression in the center of the macula where the retina is particularly thin (foveola). Within 15 minutes after death, or during life if the macula is viewed with red-free light, it appears yellow due to a carotenoid pigment (macula lutea). In this region, elongated photoreceptors, which are functionally cones, but rod-like in appearance, are abundant. The ganglion cell layer of the retina adjacent to the foveola contains several layers of large neurons, unlike the rest of the retina. The retinal pigment epithelium, which consists of a single layer of cells, extends to the optic disc posteriorly. At the ora serrata, it extends anteriorly where it fuses with the anterior continuation of the neurosensory retina and continues forward as the pigmented ciliary epithelium. Round or oval melanin granules are abundant at the apex of the retinal pigment epithelial cells. The same amount of pigment is present in all races. The retinal pigment epithelium is actively phagocytic and normally contains lipofuscin, the residual product of undigested material. The retinal pigment epithelium rests on Bruch's membrane which interdigitates with the choriocapillaris. Bruch's membrane consists of five layers: the basement membrane of the pigment epithelium, a layer of collagen, a layer of elastic tissue, a layer of collagen, and in many areas the basement membrane of the choriocapillaris. Excrescences occur in Bruch's membrane beneath the pigment epithelium (*drusen*) where they appear clinically as white dots deep to the retina. They are products of an abnormal retinal pigment epithelium.

The retina has one of the highest oxygen consumptions in the body and

202   The Eye

does not survive prolonged hypoxia. The retina can metabolize glucose to lactic acid, even under aerobic conditions.

Retinal Degenerations

A variety of retinal degenerations may be localized to part of the retina or involve extensive parts of it. The term tapetoretinal degeneration refers to degeneration of the retinal pigment epithelium and the sensory retina.

Peripheral Cystoid Degeneration

An interlacing network of interconnecting cystoid channels occurs in the peripheral retina of both eyes, especially on the temporal side of individuals over the age of 20. The condition often begins in early childhood and increases in incidence with age (Figs. 11.10, 11.11). There is severe dissolution of the retina between the inner plexiform layer and the outer limiting membrane. The cystoid spaces are separated by

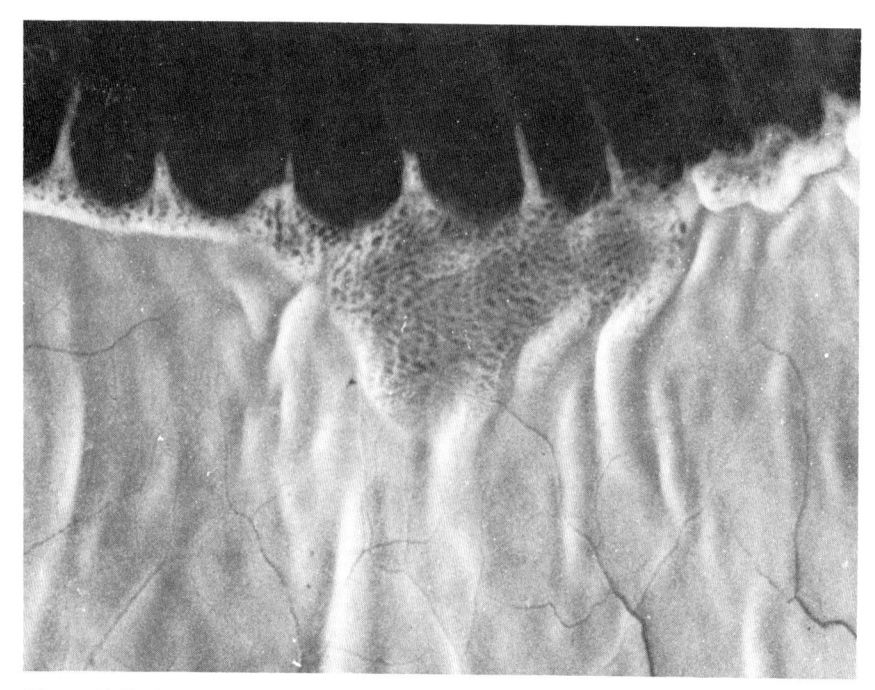

Figure 11.10. A portion of this peripheral retina immediately behind the ora serrata has a finely stippled appearance. The degree and incidence of this type of change increases with age.

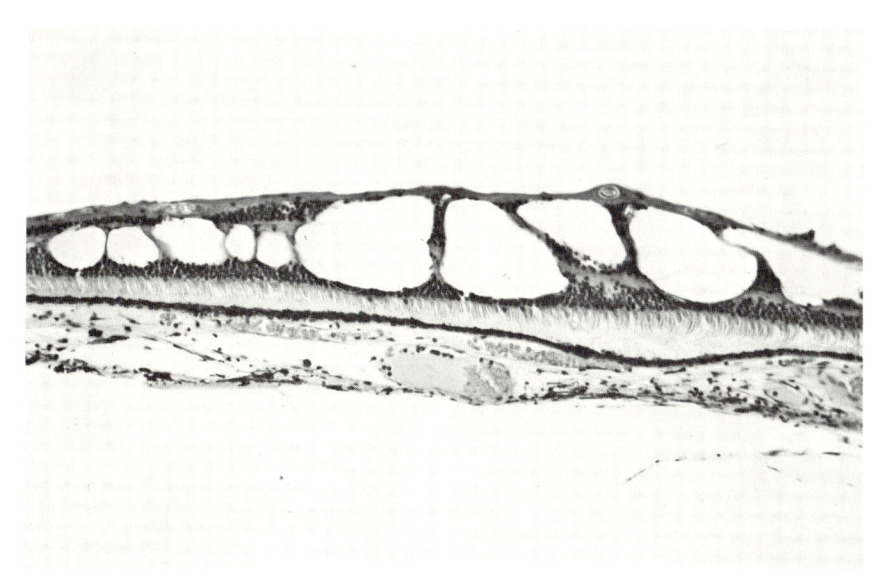

Figure 11.11. A histologic section through the stippled area shown in Fig. 11.10 reveals numerous cystoid spaces separated by radial pillars, hence the term *peripheral cystoid degeneration*. (H & E, × 100)

cellular radial pillars that bridge across the retina, giving a uniform stippled appearance when viewed from the surface (typical cystoid degeneration). The cystoid spaces contain hyaluronidase-sensitive mucopolysaccharide. The disorder begins at the ora serrata and spreads posteriorly toward the equator. Less frequently, the outer retina is preserved and delicate vascular strands form a reticulated surface pattern over the stippled internal surface of the retina (reticular cystoid degeneration of the retina).

Retinoschisis

In adult life, an exaggeration of the peripheral cystoid degeneration may be an extensive splitting of the retina, especially in its lower temporal peripheral portion (senile retinoschisis). The disruption of the radial pillars within peripheral cystoid degeneration may result in a separation of the inner and outer layers. Retinoschisis uncommonly may extend into the posterior retina, be complicated by breaks in the outer retina, and be associated with sensory retinal detachment.

Another distinctively different variety of retinoschisis (juvenile retinoschisis) occurs in infants and children and is usually inherited as an X-linked recessive disease. Splitting begins in the nerve fiber layer, often in the inferior temporal quadrant. The peripheral retina, as well as the macula, can be affected.

Lattice Degeneration

When fully developed, lattice degeneration is characterized by sharply demarcated, circumferentially oriented lesions located at or slightly anterior to the equator of the globe. The affected retina is thin. Thickened blood vessels, often with obliterated lumina, form an arborizing network of delicate white lines. In early lesions, there is a loss of the inner layers of the retina, but eventually all layers are involved. The adjacent vitreous and retinal pigment epithelium are abnormal. An abnormal vitreoretinal adhesion predisposes lattice degeneration to retinal tears. This form of degeneration occurs in about 6% of the population. It is usually bilateral and is most frequently observed in the 4th and 5th decades of life, but has been recognized over a wide age range.

Cobblestone Degeneration of the Retina

In this retinopathy, punched out areas of depigmentation and retinal atrophy are located between the equator and the ora serrata (Fig. 11.12). The thinned retina is adherent to Bruch's membrane at areas devoid of pigment epithelium. The lesions are often surrounded by a pigmented rim of proliferated retinal pigment epithelial cells and may look like chorioretinitis to the novice. The degeneration usually begins in the inferior retina close to the ora serrata and eventually may extend completely around the peripheral retina. This condition is nearly always bilateral.

Macular Diseases

A variety of disorders of the retina occur in the region of the macula. Several factors predispose to this localization, including a circulation in the choroid that differs from that of the rest of the retina, a preponderance of cones rather than rods, the nonrenewal of the cone "outer discs," and a concentration of ganglion cells. The macular disorders include central serous retinopathy and the disciform macular degenerations (Fig. 11.13). A cystoid maculopathy can occur after ocular surgery, apparently related to leaking capillaries around the macula in the retina.

"Senile" macular degeneration is a common cause of decreased central vision affecting as many as 30% of patients over age 80. Although a familial occurrence has been reported several times, many cases show no clear-cut family history of this disease, perhaps partly due to the difficulties with genetic investigations in this age group.

Pigmentary Retinopathies

A variety of conditions results in a retinopathy characterized by a loss of visual receptors (especially rods) and a focal proliferation of the

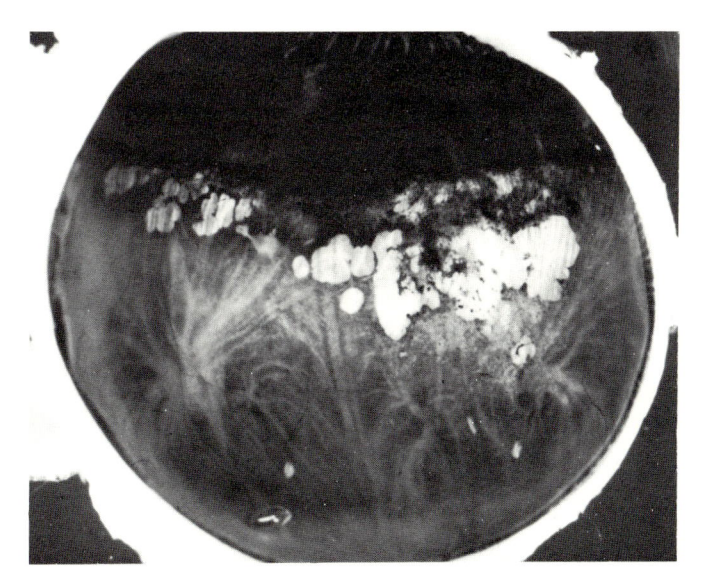

Figure 11.12. In marked contrast to the retinopathy shown in Fig. 11.10, this one in the same location is characterized by numerous punched out areas of depigmentation that resemble cobblestones in appearance.

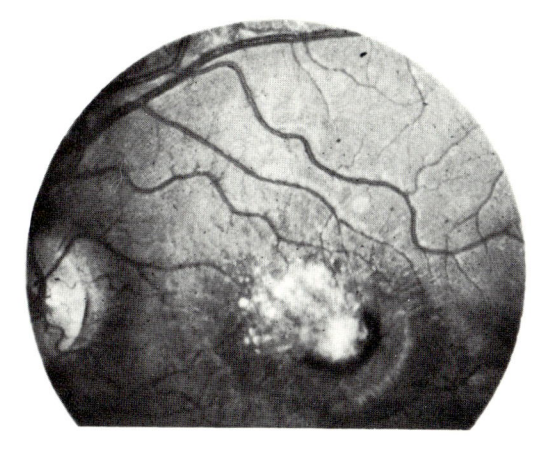

Figure 11.13. Disciform macular degeneration. An exudate is present in the macula, surrounded by a subretinal hemorrhage temporally. The retina is tented up over this lesion.

adjacent retinal pigment epithelium with migration of it into the sensory retina (pigmentary retinopathy) (Fig. 11.14, Table 11.1). The clinical manifestations, including the appearance and distribution of the retinal pigmentation, vary with the cause of the retinopathy. The pigment accumulates mainly around small branching blood vessels where the black deposits appear funduscopically like slender processes of spidery cells that have been likened to spicules of bone. A gradual attenuation of the retinal blood vessels ensues.

The term retinitis pigmentosa is often restricted to a bilateral progressive variety of pigmentary retinopathy that has a variable mode of inheritance and an onset usually in childhood. The alterations begin at the equator and the loss of rods results in night blindness, which may be an early symptom. As the condition progresses, there is a contraction of the visual fields and eventual blindness.

Sensory Retinal Detachment

The sensory retina can become separated from the retinal pigment epithelium by liquid vitreous, hemorrhage, neoplasm, and/or exudate. Retinal detachment is bilateral about 20–25% of the time. Retinal holes and weakening of the fixation of the retina predispose to sensory retinal detachment. Full thickness holes in the retina are common, but are usually not complicated by retinal detachment. However, if the hole allows liquid vitreous to gain access to the potential space between the retina and the retinal pigment epithelium, these layers often separate. A

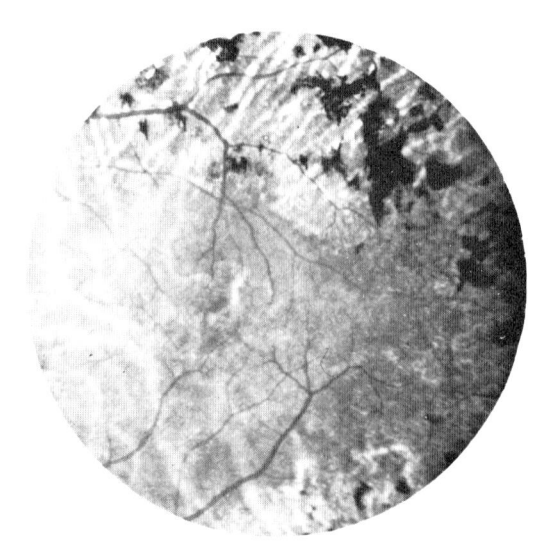

Figure 11.14. Pigmentary retinopathy. There is atrophy of the retinal pigment epithelium with unmasking of the underlying choroidal vascular pattern. Pigment migration and clumping is seen, in large clumps and also in stellate-shaped clumps which are related to blood vessels.

TABLE 11.1
Some Conditions Associated with a Pigmentary Retinopathy

1. Retinitis pigmentosa
2. Postinflammatory and degenerative:
    Syphilis
    Typhoid fever
    Rubella
    Measles
    Smallpox vaccination
    Behcet's disease
    Cytomegalic inclusion disease
    Toxoplasmosis
    Onchocerciasis
3. Drugs:
    Phenothiazine: chlorpromazine, iminophenoxine
    Antimalarials: chloroquine, atabrine
    Antimetabolites: iodoacetate, sodium fluoride
4. Disorders of lipid and mucopolysaccharide metabolism
    Neuronal ceroid lipofuscinosis
    Bassen-Kornzweig (abetalipoproteinemia)
    Refsum syndrome
    Niemann-Pick disease
    Hurler Syndrome (MPS 1-H)
    Scheie syndrome (MPS 1-S)
    Hunter syndrome (MPS II)
    Sanfilippo syndromes A and B (MPS III)
5. Miscellaneous syndromes:
    Kearn-Sayre
    Usher
    Cockayne
    Laurence-Moon
    Bardet-Biedl
    Pelizaeus-Merzbacher
    Hallervorden-Spatz
    Siderosis bulbi
    Congenital ishthyosis
    Hallgren
    Vitamin A deficiency

liquid vitreous and even some vitreoretinal traction seem to be necessary for retinal detachment to occur. Retinal holes often stem from trauma and specific retinal degenerations. The retina may be pulled inward by inherent vitreoretinal adhesions, as occurs with retinitis proliferans, retinopathy of permaturity, and after intraocular infection. Vitreous traction may lead to tears in the retina and retinal detachment. The vitreous body normally maintains the retina in its normal position. A diminution in the pressure exerted by the vitreous, as after vitreous loss, also predisposes to sensory retinal detachment.

Retinal detachment is common after physical trauma to the eye and may result from intraocular hemorrhage. It can also follow retinal tears, a macular hole, or a disinsertion of the retina at the ora serrata which permits vitreous to pass between the sensory retina and the retinal pigment epithelium. Detached retinas may occur 20 years after removal of congenital cataracts.

Irrespective of the cause of the retinal detachment, a protein-rich subretinal fluid accumulates and the outer portion of the retina degenerates because of the loss of a crucial functional relationship with the retinal pigment epithelium. A spatial separation of the photoreceptors and retinal pigment epithelium interferes with the metabolic interaction between these two retinal layers. Moreover, nutrients that normally reach the outer retina from the choriocapillaris need to diffuse a greater distance. After the sensory retina has been detached for a long time, adhesions develop between adjacent portions of the convoluted retina. The shriveled-up retina passes from the optic disc to the back of the lens, which is commonly cataractous (Fig. 11.15). Sometimes, pseudocysts form in the retina. An excessive proliferation of the ciliary and retinal pigment epithelium may encircle the globe at the ora serrata (*ringschwiele*) (Fig. 11.16). Characteristic, fine structural changes accompany sensory retinal detachment. The outer segments of the photoreceptors degenerate. Cystoid extracellular spaces appear between the middle and inner layers of the retina. Melanin granules withdraw from the apical processes of the retinal pigment epithelium and lamellar inclusion bodies disappear from their cytoplasm.

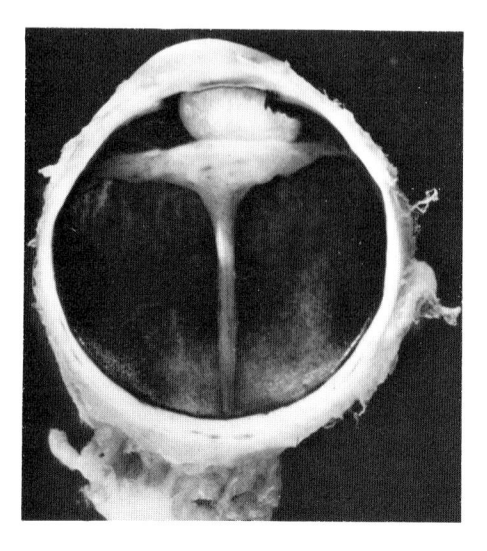

Figure 11.15. This eye has a chronic sensory retinal detachment. Note how the retina remains attached to the ora serrata and optic disc.

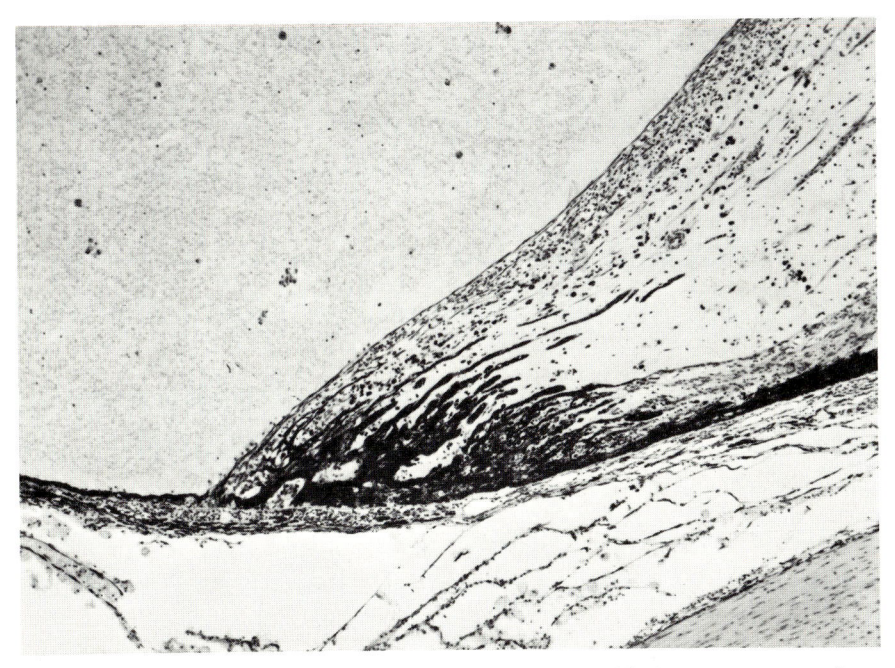

Figure 11.16. The retinal pigment epithelium proliferates readily in response to a variety of stimuli. With chronic sensory retinal detachment, a zone of retinal pigment epithelium proliferation often encircles the globe at the ora serrata (ringschwiele). (H & E, × 50)

LENS

A single layer of cuboidal cells is present immediately beneath the anterior surface of the lens capsule, but not normally beneath the posterior capsule. At the equator of the lens, the cuboidal cells differentiate into elongated lens fibers, which interdigitate with each other with insignificant space between the individual fibers. The lens has no blood supply and almost no mitochondria. It contains specific proteins that do not exist elsewhere in the body (soluble α, β, and γ crystallins and an insoluble albuminoid). With aging, the insoluble albuminoid, derived from α crystallin, as well as a yellow pigment increase in quantity. The crystalline lens increases in size with age. The lens is elastic and has a tendency to become spherical. A loss of this elasticity results in presbyopia.

Cataracts

Opacities in the crystalline lens are termed cataracts (Fig. 11.17). They are a major cause of visual impairment and blindness throughout

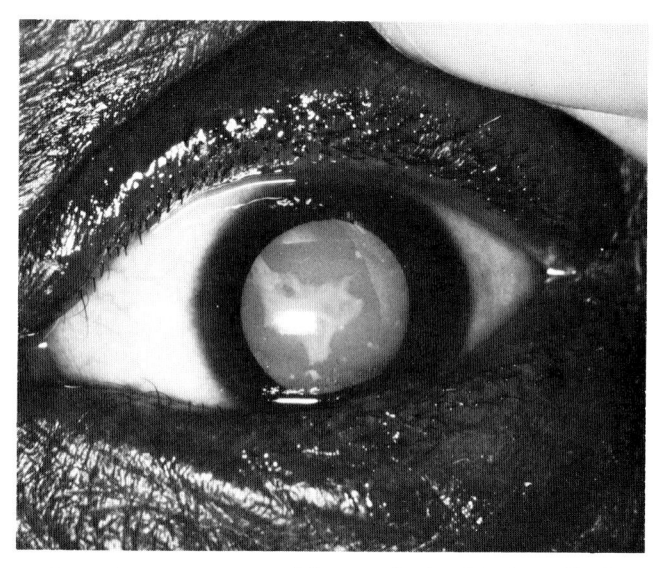

Figure 11.17. A mature cataract. The completely white lens fills the pupil.

the world. Cataracts have many causes (Table 11.2). Different types of cataracts develop as individuals become older and are the rule in advanced age. In the most common "senile" cataracts, wedge- or spoke-like opacities occur at the lens equator and extend into the anterior and posterior cortex (*cortical cataract*). The cortical cataract becomes first clinically evident with the slit lamp, because vacuoles appear in the cortex of the lens. Later, radial clefts form which become filled with opaque debris. In another type of cataract, granular flaky opacities gradually appear in the zone immediately anterior to the posterior capsule and spread toward the periphery (*posterior subcapsular cataract*). Clefts appear between the lens fibers of the cortex and contain degenerated lens material (*Morgagnian corpuscles, incipient cataract*). A lens with an immature cataract can increase in volume by imbibing water as a result of the osmotic pressure exerted by the degenerated lens material ("*intumescent cataract*"). The swollen lens may obstruct the pupil and cause glaucoma. Eventually, the entire lens becomes affected by the degenerative process ("*mature cataract*"). Such cataractous lenses subsequently shrink after the lens capsule and epithelium degenerate and lenticular debris escapes into the aqueous humor ("*hypermature cataract*"). In many cases, the lens capsule is presumed to be intact, but sufficiently permeable to allow lens material to escape. The extruded lenticular debris becomes engulfed by macrophages which may obstruct the aqueous outflow resulting in phacolytic glaucoma. The compressed lens fibers in the central portion of the lens normally harden with aging (*simple nuclear sclerotic cataract*). An intensification of this process opacifies the affected area which may become discolored to a brown or even black color. The sclerotic nucleus may sink in the lens if the lens cortex becomes entirely liquified (*Morgagnian cataract*). After an extracapsular cataract extraction, and rarely in the absence of surgery, there is reabsorption of all lens proteins leaving only a clear lens capsule. Cataractous lenses commonly calcify.

If, after the surgical removal of a cataractous lens, the capsular epithelium or other nucleated lens fibers remain within the eye, these cells may proliferate and produce abortive lens fibers. These sometimes appear as large globules. (*Elschnig's pearls*). Their formation can also follow ocular trauma. After the loss of the internal contents of the lens, either as a result of cataract surgery or trauma, the anterior and posterior capsules of the lens may become apposed while the new lens fibers form. This results in a doughnut-shaped lens (*Soemmering's ring cataract*).

A proliferation of the epithelium beneath the lens capsule is common in some congenital cataracts and in other cataracts due to iritis, electric shock, and trauma. This epithelium may migrate and extend over the

TABLE 11.2
Some Causes of Cataracts

A. Metabolic:
Disorders of carbohydrate metabolism: monosaccharide excesses (e.g. galactose, zylose, glucose), galactosemia, diabetes mellitus
Endocrine disorders: cretinism
Chromosomal anomalies: Down's syndrome
Disorders of calcium metabolism: hypocalcemia
Deficiency states: riboflavin, tryptophan
Genetic disorders: myotonia dystrophica, numerous inherited cataracts, Fabry's disease
Toxin: dinitrophenol, naphthalene, thalium, myleran, ergot, Mer-29, miotics, phenothiazines, steroids
B. Physical agents:
Heat, radiation, trauma, ultrasound, electric current
C. Ocular diseases:
Inflammation of eye (e.g. sarcoidosis), retinitis pigmentosa, myopia, intraocular neoplasms, glaucoma, sensory retinal detachment
D. Viruses:
Rubella
E. Aging
F. Skin disease:
Atopic eczema, scleroderma

posterior lens capsule. Some epithelial cells may enlarge and appear vacuolated in tissue sections (*bladder cells*). This is common in cataracts due to choroiditis and other diseases in the posterior segment of the eye and is occasionally seen in senile cataracts. The epithelium of the lens may undergo necrosis in association with: acute glaucoma, hyphema, postoperative anterior segment necrosis, chemical injury, and advanced senile cataracts. Anterior polar cataracts, with subepithelial plaques containing abortive lens fibers and collagen, can form. A retention of nuclei in the fibers of the center of the lens commonly occurs in cataracts due to rubella, trisomy 13, and other conditions. Excrescences on the lens capsule occur in Down's syndrome.

Displaced Lenses

The zonular fibers which connect the nonpigmented epithelium of the ciliary body with the lens consist of densely packed filaments with a 200 Å periodicity. Should these zonules, or their attachments to the lens, be defective as a result of a developmental anomaly, trauma, or disease, the lens becomes displaced from its normal location. If some zonules remain intact, the lens may stay in the posterior chamber behind the iris in a saucer-shaped depression in the anterior surface of the vitreous body (*subluxation*), or it may by displaced completely out of its normal position (*dislocation*). The lens can become displaced posteriorly with prolapse of the vitreous into the pupil; an ectopic lens may also become located in the anterior chamber. Ectopic lenses occur with spherophakia, homocystinuria, the syndromes of Weill-Marchesani and Marfan, and in other rare systemic conditions. A progressive downward displacement is more common in homocystinuria than in Marfan's syndrome. The iris lacks support when the lens is displaced and is tremulous on ocular movement (*iridodonesis*). Glaucoma commonly accompanies dislocated lenses.

Exfoliation and Pseudoexfoliation of the Lens Capsule

The crystalline lens is covered by a hyaline capsule which is the thickest basement membrane in the body. It thickens with age, but it remains thinner posteriorly throughout life. True exfoliation of the lens capsule is rare. In the past, it occurred mainly in workers exposed to intense heat over many years. The exfoliation is usually bilateral, affects the region within the pupil, and is thought to be due to infrared radiation. A condition, characterized by unilateral or bilateral deposits on the lens capsule, zonules, ciliary body, iris, and in the angle of the anterior chamber, has been designated pseudoexfoliation of the lens capsule. The deposits involve particularly the pupillary area. The origin of the material is still disputed. It possesses some staining characteristics of zonules, but differs in its filamentous ultrastructure. Some analyses have contained amyloid-like material. The iris develops a moth-eaten appearance, due to an associated degeneration of its pigment epithelium. A dense pigmentation of the anterior chamber angle is common. Cataracts and an open angle glaucoma are frequently associated.

VITREOUS BODY

The vitreous body is a transparent, reticulated gel behind the lens. Its central portion contains delicate collagen fibrils, as well as much hyaluronic acid and water. Filaments are densely packed anteriorly and adherent to the posterior surface of the lens (*Weigert's hyaloid-capsular ligament*), particularly in young people. Normally, the vitreous is firmly adherent to the inner limiting membrane of the retina at the ora serrata (vitreal base), optic disc, and probably around the macula. Abnormal vitreoretinal adhesions occur over chorioretinal scars, lattice retinal degeneration, and along blood vessels with age. Although the vitreous is essentially acellular, a few flattened hyalocytes are present, especially in the peripheral cortical part.

Vitreal Opacities

Intact or degenerating cells, such as leukocytes, macrophages, erythrocytes, ciliary epithelium, or neoplastic cells appear in the normally crystal clear vitreous as opaque particles. Their presence within the vitreous may interfere with the transmission of light to the retina and some become visible to the living individual as small defects in the visual field. The opacities may also be detectable on ophthalmoscopic examination of the vitreous body.

Asteroid Hyalitis

Occasionally, in eyes free from overt disease, innumerable small spherical or disc-shaped white or creamy opacities appear in the vitreous (Figs. 11.18, 11.19). This asymptomatic entity occurs especially in the elderly and is usually unilateral (75% of cases). There is no tendency for the deposits to settle with gravity. The precise nature of the opacities still needs to be established. They stain positively with histochemical techniques for calcium, mucopolysaccharides, and lipids, but resist lipid solvents. There is usually no cellular infiltrate, but the normal vitreal opacities may be surrounded by multinucleated foreign body giant cells.

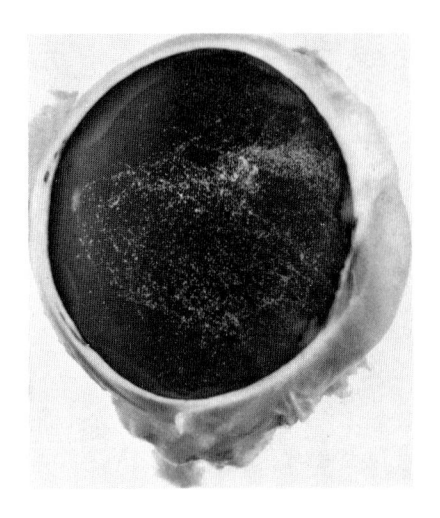

Figure 11.18. Numerous small opacities are scattered throughout the vitreous in this eye with asteroid hyalitis.

Synchisis Scintillans (Cholesterolosis Bulbi)

Shiny, angular, rectangular, crystalline vitreal opacities may be associated with chronic intraocular diseases. Unlike the more common asteroid hyalitis, the crystalline sediment sinks to the dependent parts of the vitreous, reappears on movement, and usually involves both eyes. Most patients with synchisis scintillans are over 35 years of age. The deposits are thought to consist of cholesterol, fatty acids, and calcium phosphate.

Liquefaction of Vitreous Body (Syneresis)

With age, clear areas (cavities) may develop in the vitreous. These fluid-filled spaces enlarge and rupture through the vitreoretinal interface with collapse of the remaining vitreous gel. This may exert traction on the retina at its points of attachment and may initiate nerve impulses, interpreted by the patient as flashing lights. Small blood vessels can be torn, producing bleeding into the vitreous cavity, and are noted by the patient as "floaters." Ocular inflammation, trauma, and surgery can accelerate this process of liquefaction of the vitreous.

OPTIC NERVE

After the nerve fibers from the retina pass through the lamina cribrosa, they normally become myelinated. The optic nerve is surrounded by meningeal sheaths of dura, arachnoid, and pia mater, and it contains neuroglia (oligodendrocytes, astrocytes, and microglia) like those in

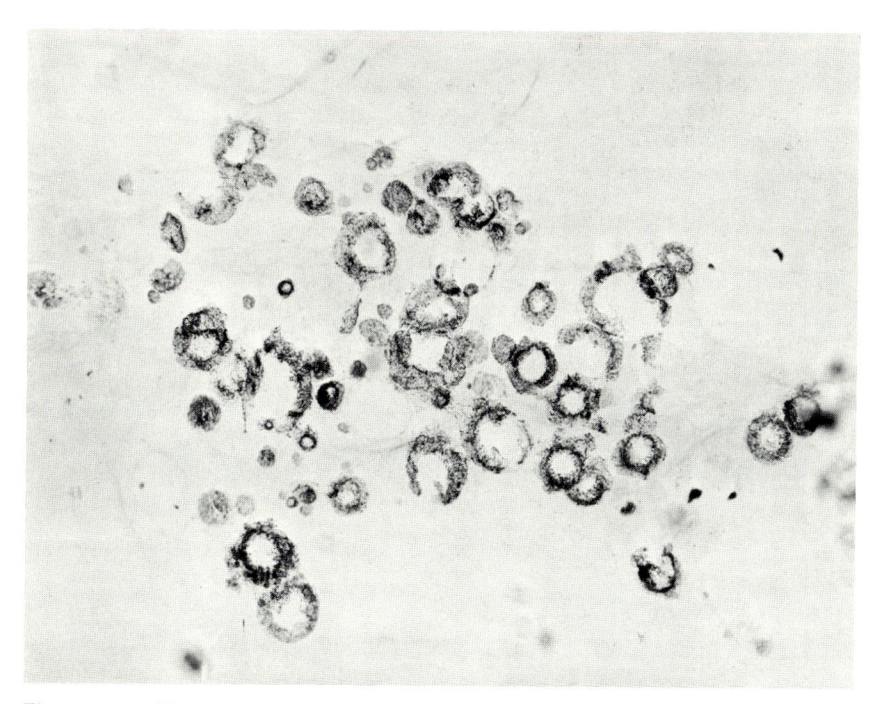

Figure 11.19. The opacities in asteroid hyalitis stain with several histochemical techniques. (Hale's colloidal iron, × 100)

other parts of the central nervous system. Fibrovascular septa extend from the pia between the myelinated axons, a feature different from other tracts of the central nervous system.

Optic Atrophy

Optic atrophy follows a loss of axons and their myelin sheaths, within the optic nerve. Any condition which damages the retinal ganglion cells or their axons within the retina or the optic nerve can cause optic atrophy (Fig. 11.20). Several varieties of optic atrophy with different modes of inheritance occur. Optic atrophy can follow a lesion in the retina (ascending optic atrophy) or one that is proximal to the eye (descending optic atrophy). Discrete retinal lesions will affect only that portion of the optic nerve containing axons derived from, or passing across, that part of the retina. For example, lesions in the macula affect the papillo-macular bundle in the optic nerve. The appearance of the optic nerve varies with the degree of optic atrophy and its cause. In descending optic atrophy, the changes at the disc depend on the proximity of the primary site of

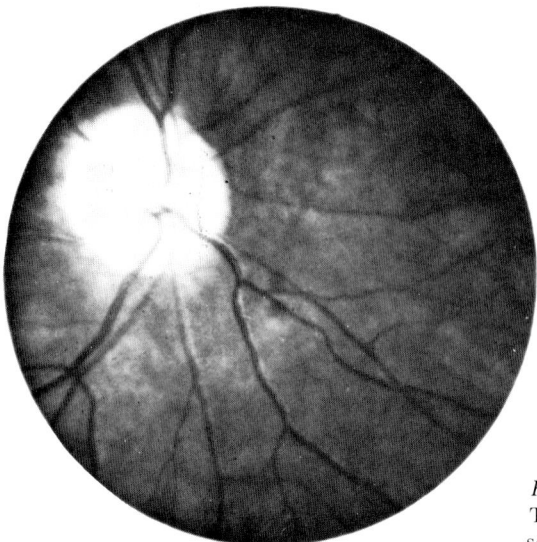

Figure 11.20. Optic atrophy. The nerve head is pale with absence of small blood vessels.

nerve damage to the eye, the nature of the initial process, and the duration of the lesion. The optic disc may initially be pale pink to gray, but eventually becomes white with loss of capillaries which nourish the optic nerve. The margins of the disc are discrete and the lamina cribrosa mottled.

Aside from the diminution in the amount of parenchymatous tissue due to the loss of axons and myelin, there is often a widening of the subarachnoid space and redundancy of dura mater. Particularly when optic atrophy follows a chronic lesion of the intracranial portion of the optic nerve, the physiologic cup within the nerve head may be widened and deepened. In other instances, it may be filled in. There is generally a minimal proliferation of astrocytes and an insignificant increase in connective tissue of the pial septa when the primary lesion is intracranial. However, when the optic atrophy follows an inflammatory or vascular lesion of the nerve, pronounced reactive alterations in the glial and mesenchymal tissues of the nerve frequently occur. An unusual variety of optic atrophy is characterized by cavernous spaces filled with mucoid material (*cavernous optic atrophy*). It is nearly always associated with glaucoma.

EYELIDS

The eyelids contain striated muscle (levator palpebrae superioris, and orbicularis oculi), smooth muscle (Müller's muscle), lymphatics, eyelashes, hair follicles, modified sweat glands (Moll's glands), sebaceous glands, and accessory lacrimal glands of Krause and Wolfring, but normally not adipose connective tissue. The upper lid possesses a dense connective tissue (the tarsal plates), within which are located the Meibomian glands.

With chronic inflammation of the conjunctiva or cornea and occasionally after ocular surgery, an excessive contraction of the orbicularis oculi can occur (*blepharospasm*). A loss of the tonicity of the orbicularis oculi as in the aged or with scarring of the eyelids may cause the eyelids to be turned inward. Entropion leads to the irritation of the cornea and conjunctiva by the eyelashes.

Inadequate closure of the eyelids results from severe proptosis, an enlarged globe, senile relaxation of the lower lid, and excessive surgery for blepharoptosis or lid retraction, as well as from facial nerve palsies in which the function of the orbicularis oculi muscle is defective (Fig. 11.21). Scarring of the eyelids can cause the upper and lower lids to turn away from the globe. An eversion of the eyelid (*ectropion*) or failure of lids to close and cover the eye (*lagophthalmos*) exposes the conjunctiva and cornea, often resulting in exposure keratopathy.

ORBIT

The orbit also contains many structures which may become diseased. These include adipose, elastic, and fibrous connective tissues, blood vessels, nerves, and sympathetic ganglia, as well as smooth and skeletal

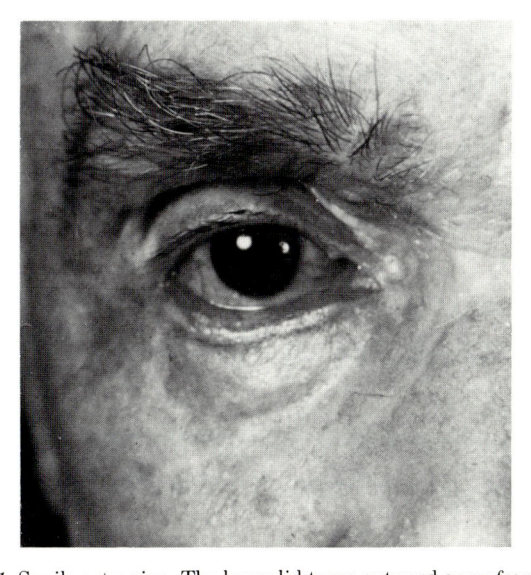

Figure 11.21. Senile ectropion. The lower lid turns outward away from the globe.

TABLE 11.3

Some Conditions other than Orbital Neoplasms the Cause Exophthalmos

A. Metabolic
    Hyperthyroidism
B. Orbital conditions
    Inflammatory lesions:
        Nodular fasciitis
        Orbital abscess
        Foreign body granuloma
        Wegener's granulomatosis
        Orbital cellulitis
        Histiocytosis-X
        Nonspecific idiopathic inflammation (pseudotumors)
        Sarcoidosis
        Dacryoadenitis
        Posterior scleritis
        Myositis
        Mucormycosis
        Aspergillosis
        Onchocerciasis
    Developmental anomalies:
        Encephalocele
        Meningo-encephalocele
        Hemangiomas
        Lymphangiomas
        Teratoma
        Ectopic brain tissue
        Dermoid cysts
    Vascular:
        Retrobulbar hematoma
        Aneurysm of ophthalmic artery
    Miscellaneous conditions:
        Xanthomatosis of orbital fat
        Anomalies of bony orbit
        Orbital emphysema
        Epidermoid implantation cysts
        Amyloid
        Parasitic cysts
C. Lesions of paranasal sinuses
    Mucocele
    Empyema
    Carcinoma
D. Intracranial lesions
    Cavernous sinus thrombosis
    Carotid-cavernous fistula
    Meningioma
    Hydrocephalus

muscles. The trochlea of the superior oblique muscle is the sole cartilaginous structure present in the orbit. The lacrimal gland, the only epithelial tissue in the orbit, has a structure similar to the salivary glands.

An abnormal protrusion of the eyeball from the orbit may result from many causes (Fig. 3.1, Table 11.3). Exposure keratopathy may be a sequel to severe proptosis. The degree of proptosis depends on the size, nature, and position of the lesion in the orbit. Conditions which may simulate a protruded globe include extreme myopia, asymmetric bony orbits, buphthalmos, unilateral upper lid retraction, and relaxation of one or more rectus muscles.

MYOPIA

Myopia is a refractive ocular abnormality in which light from the visualized object is brought to focus at a point in front of the retina. This may be due to the anteroposterior diameter of the eye being longer than usual or to the lens being excessively strong. Myopia usually has its onset in youth and mild and severe forms are recognized. The mild form (stationary or simple myopia) is generally nonprogressive after the cessation of body growth and is less severe than the genetically determined "progressive myopia." A thin sclera, especially at the posterior pole, an atrophic choroid, tears in Bruch's membrane, and flattening of the optic disc on the temporal side may occur in marked myopia. In severe myopia, the retinal pigment epithelium often stops short of the disc and the bare sclera or choroid can be seen through the retina, usually at the temporal edge of the optic disc. Open angle glaucoma and sensory retinal detachment are more common in extreme degrees of myopia than in the general population.

REFERENCES

ASHTON, N., SHAKIB, M., COLLYER, R., AND BLACH, R. Electron microscopic study of pseudo-exfoliation of the lens capsule. I. Lens capsule and zonular fibers. Invest. Ophthalmol. 4:141–153, 1965.

ELLIOT, K., AND FITZSIMONS, D. W. (Eds). The Human Lens in Relation to Cataract. Associated Scientific Publishers, Amsterdam, 1973.

FARKAS, T. G., SYLVESTER, V., AND ARCHER, D. The ultrastructure of drusen. Am. J. Ophthalmol. 71:1196–1205, 1971.

FARKAS, T. G., SYLVESTER, V., AND ARCHER, D. The histochemistry of drusen. Am. J. Ophthalmol. 71:1206–1215, 1971.

Foos, R. Y., AND FEMAN, S. S. Reticular cystoid degeneration of the peripheral retina. Am. J. Ophthalmol. 69:392–403, 1970.

GASS, J. D. M. Pathogenesis of disciform detachment of the neuroepithelium. Am. J. Ophthalmol. 63:573–711, 1967.

GOLDMAN, J. N., AND BENEDEK, G. B. The relationship between morphology and transparency in the nonswelling corneal stroma of the shark. Invest. Ophthalmol. 6:574–600, 1967.

JAFFE, N. S. The Vitreous in Clinical Ophthalmology. C. V. Mosby, Co., St. Louis, 1969.

MACHEMER, R., AND KROLL, A. J. Experimental retinal detachment in the owl monkey. VII. Photoreceptor protein in normal and detached retina. Am. J. Ophthalmol. 71:690–695, 1971.

SHAKIB, M., ASHTON, N., AND BLACH, R. Electron microscopic study of pseudo-exfoliation of the lens capsule. II. Iris and ciliary body. Invest. Ophthalmol. 4:154–161, 1965.

STRAATSMA, B. R., AND FOOS, R. Y. Typical and reticular degenerative retinoschisis. Am. J. Ophthalmol. 75:551–575, 1973.

YOUNG, R. W. The renewal of photoreceptor outer segments. J. Cell Biol. 33:61–72, 1972.

INDEX

Abetalipoproteinemia, 141, 207
Ablepharon, 83
Abscess, 26, 115, 133, 218
　orbital, 218
　vitreal, 26, 133
Acanthamoeba, 132
Acanthosis, 3
Adenoacanthoma, 90
Adenocarcinoma, 90, 96, 97, 99, 100
　ciliary epithelium, 90, 96, 97
　lacrimal gland, 100
　lacrimal sac, 101
　Meibomian gland, 99
　sweat gland, 90
Adenocystic carcinoma, 90, 100
Adenoid cystic carcinoma, 100
Adenoma, 90, 96
　accessory lacrimal glands of Krause and Wolfring, 90
　sebaceous glands (Meibomian glands), 90
　sweat glands and apocrine glands (Moll's glands), 90
Adhesions, 207, 213
　vitreoretinal, 207, 213
Adipose connective tissue, 1
Adrenal insufficiency, 163
Agranulocytosis, 114
Albinism, 139, 151, 152
Alkaptonuria, 139, 149, 150
Allergic reactions to
　adhesive tapes, 33
　cosmetics, 33
　hair sprays, 33
　microorganisms, 33
　nail polishes, 33
　pollens, 33
　soaps, 33
Allergy, 33
Allograft, 34, 39
Ameba, 132, 137
Aminoacidopathies, 139
Amyloid, 10, 11, 218
Amyloidosis, 14, 15
　inherited, 11, 15
　primary heredofamilial, 143
Anaphylactogenic reaction, 33
Anemia, 48, 52
Anencephaly, 68, 87
Aneurysm
　carotid artery, 67
　ophthalmic artery, 218

Aneurysmal bone cyst, 91
Angiitis
　hypersensitivity, 42
　necrotizing, 29
Angioid streaks, 154, 198
Angioma (*see also* Hemangioma)
　retina, 57
Angiomatosis
　encephalofacial, 86
　encephalotrigeminal, 86
Aniridia, 78, 79, 87, 186
Ankyloblepharon, 83
Ankylosing spondylitis, 24, 25
Anophthalmia, 68
　primary, 68
　secondary, 68
Aphakia, 70
Aplasia, 70
Aqueous flare, 17, 24
Arcus senilis, 8, 9
Argyrosis, 14
Armanni-Epstein phenomenon, 146
Arterial sheathing, 66
Arteriolitis, necrotizing, 66
Arteriolosclerosis, 66
Arteriosclerosis, 58, 64
Arteriovenous fistulae, 67
Arteriovenous malformations, 84
Arteriovenous nicking, 66
Arteritis
　cranial, 66
　giant cell, 60, 66, 189
　temporal, 66, 67
Arthritis, 29
Arthropods, 126, 136
Arthus reactions, 35, 36, 43
Aspergillosis, 127
Asteroid hyalitis, 213–215
Atabrine, 207
Ataxia-telangiectasia, 143
Atheromatous plaques, 60
Atherosclerosis, 65, 189
Atopic eczema, 211
Atopic reactions, 33
Atrial myxoma, 60
Atrophy, 1
Atropine, 22
Autoallergies, 37

Bacillary dysentery, 29, 117
Bacterial endocarditis, 60, 116

Barometric decompression, 61
Basal cell carcinoma, 90, 98
Basal cell papilloma, 98
Benign mucous membrane pemphigoid, 23, 195
Bergmeister papilla, 82
Bitot spots, 162
Blastomyces dermatitidis, 127
Bleeding diathesis, 46
Blepharitis, 16, 115
  staphylococcal, 115
Blepharospasm, 217
Blood retinal barrier, 46
Bodies
  Birbeck, 30
  foreign, 19, 114, 136, 170, 172
  Guarnieri, 125
  Halberstaedter-Prowazek, 129
  Hassal-Henle, 10
  Henderson-Patterson, 123
  inclusion, 22, 120, 121, 123, 129
  Lipschütz, 120, 123
  metallic, foreign, 170
  Molluscum, 123
  Morgagnian, 210
  Russell, 6
Boils, 115
Bone, 1, 2
Brachycephaly, 85
Bronze, 14, 156
Brucellosis, 24, 117
Brushfield spots, 87
Buphthalmos, 81, 183, 184, 186, 219
Burns, 28, 173, 176, 177
  alkali, 177, 178, 179
  caustic, 23
  chemical, 22, 28
Busacca nodules, 18

Café-au-lait spots, 7, 158
Calabar swelling, 135
Calcification, 12, 13, 71, 94, 96, 97, 121, 133, 159
Calcifying epithelioma of Malherbe, 90
Calcium, 11, 12, 156, 162
Calliphoridae, 136
Candida, 127, 137
Candida albicans, 127
Carbuncles, 115
Carcinoma, 89, 90, 98, 108, 109, 112, 218
  basal cell, 90, 98
  intraepithelial (in situ), 90, 99
  Meibomian gland, 90
  metastatic, 108, 109, 112
  mucoepidermoid, 90, 101
  paranasal sinuses, 218
  sebaceous, 112
  squamous cell, 90, 97, 99–101, 112

Carotene, 161
Cartilage, 1, 70, 77, 87, 96
Cat scratch fever, 29, 115
Cat's eye reflex, 92, 93
Cataract, 1, 7, 11, 13, 14, 26, 39, 78, 79, 81, 87, 91, 97, 107, 119, 123, 147, 148, 149, 152, 153, 156, 164, 168, 169, 170, 171, 174, 176, 177, 188, 208–213, 219
  anterior polar, 212
  anterior subcapsular, 168
  calcified, 1, 13
  causes, 211
  congenital, 78, 87
  cortical, 210
  diabetic, 147
  extraction, 169, 170, 188, 211
  furnaceman's, 174
  galactosemia, 148, 149
  glassblower's, 174
  hypermature, 188, 210
  incipient, 210
  industrial heat, 174
  infrared, 174
  intumescent, 188, 210
  Lowe's oculocerebrorenal syndrome, 153
  mature, 210
  Morgagnian, 211
  posterior polar, 78, 79
  posterior subcapsular, 25, 164, 210
  radiation, 174
  rubella, 78, 212
  senile, 147, 212
  simple nuclear sclerotic, 211
  snowflake, 147
  Soemmering's ring, 211
  subcapsular, 171
  sunflower, 14, 156
  surgery, 169, 170, 188, 211
Caterpillars, 136
Cavernous sinus thrombosis, 45, 52, 115
Cells
  B, 35
  bladder, 212
  eosinophils, 22, 33, 39, 41
  epithelioid, 25, 39, 41
  foreign body giant, 39, 57, 213
  giant, 28, 31, 38, 39, 41, 57, 66, 213
  histiocytes, 20, 31, 38, 129
  Langerhans, 30
  Langhans, 30, 39
  leukocytes, 26, 67
  lymphocytes, 20, 22, 39, 40, 41
  macrophages, 18, 25, 39, 57, 58, 64, 114, 129
  monocytes, 18
  mononuclear, 20, 24
  multinucleated epithelial, 22
  multinucleated giant, 28, 31, 39, 66, 213

neutrophils, 19, 39
pericytes, retinal, 45
plasma cells, 20, 22, 35, 39, 41
polymorphonuclear leucocytes, 36, 114
T, 34
Touton giant cells, 31
Cell-mediated hypersensitivity, 20
Cell-mediated immunity, 33, 114, 120
Cellular hypersensitivity, 34
Cellulitis, orbital, 16, 107, 218
Central retinal artery, occlusion, 164
Central retinal vein, occlusion, 48, 51, 64, 67
Central retinal vein, thrombosis, 57, 154
Central serous retinopathy, 204
Cercoma, 134
Cerebroside lipidosis, 141
Chagoma, 131
Chalazion, 26, 27, 99
Chalcosis, 14, 171
Chancroid, 117
Chemosis, 56, 117, 122
Cherry red spot, 6, 64, 155, 164
Chickenpox, 119, 121
Chlamydia, 22, 128
Chlamydia oculogenitalis, 114, 129
Chlamydia trachomatis, 128
Chlorolabe, 159
Chloroquine, 163, 207
Chlorpromazine, 163, 207
Cholesterol, 11, 31, 56, 57, 60, 73, 214
Cholesterolosis bulbi, 214
Chondroma, 91
Chondrosarcoma, 89, 91
Chorioretinal gyrate atrophy, 140
Chorioretinal scar, 213
Chorioretinitis, 16, 26, 117, 119, 121, 126, 131
  syphilitic, 126
Choristoma, 70
Choroid
  choroiditis 16, 24
  melanoma, 106
  nevus, 75
Choroideremia, 144
Choroiditis, 16, 24
Chromosomal abnormalities, 1, 69, 70, 82, 86, 87, 88, 92, 113, 138, 164, 186, 212
Chromosomal anomalies
  chromosome No. 4 short arm deletion, 86
  chromosome No. 5 short arm deletion (cri-du-chat), 86
  chromosome No. 18 deletion, 1, 88
  chromosome No. 21 long arm deletion, 86
  D-group depletion syndromes, 92, 113
  partial monosomy 18, 86
  penta-X, 86
  trisomy 13, 1, 69, 70, 82, 86, 87, 164, 186, 212
  trisomy 18, 86, 87, 186
  trisomy 21, 86
  XXXXY, 86
Ciliary body
  adenocarcinoma, 90, 96, 97
  cyclodialysis, 167, 170
  cysts, 111, 112
Clostridium, 8, 118
Clostridium botulinum, 118
Clostridium perfringens, 8, 118
Clostridium tetani, 118
Coats lesion, 56, 57
Coccidiodomycosis, 115
Coccidioides immitis, 127
Collagen
  fibrous long spacing, 12
  segment long spacing, 12
  tropocollagen, 12
Coloboma, 69, 71, 73, 81, 83, 85, 86, 87, 88, 144, 186
  atypical, 69
  eyelid, 73, 83, 85
  inherited, 144
Colobomatous cyst, 69
Color blindness, 159
Commotio retinae, 168
Cone monochromatism, 159
Congenital cystic eye, 68
Congenital ichthyosis, 207
Congenital pit of the optic nerve head, 87, 88
Congenital stationary night blindness, 144
Conjunctiva
  acquired conjunctival melanosis, 7, 90, 108
  amyloid, 10, 11
  benign mucous membrane pemphigoid, 23, 195
  Bitot spots, 162
  burns, 22, 23
  chemosis, 56, 117, 122
  conjunctivitis (*see* Conjunctivitis)
  follicular reaction, 22
  hemorrhage, 47, 179
  inclusion blennorrhea, 129
  lymphangiectasia, 72
  melanoma, 109
  nevus, 75, 109
  pinguecula, 12, 14, 192-194
  pseudopterygium, 24
  pterygium, 14, 193, 195
  recurrent juvenile conjunctival papillomatosis, 99
  spring catarrh, 34
Conjunctivitis, 10, 16, 21, 22, 23, 28, 31, 34, 114, 115, 116, 117, 118, 119, 122, 123, 125, 128, 129, 136
  acute, 21

Index

Conjunctivitis—*continued*
  acute allergic, 34
  acute bacterial, 22
  allergic, 22, 34
  anaphylactogenic, 34
  catarrhal, 125
  chronic, 10, 23
  chronic recurrent, 23
  cicatrizing, 16
  epidemic Koch-Weeks, 137
  follicular, 22, 119, 122, 123, 129
  glare, 175
  granulomatous, 136
  inclusion, 22, 31, 128, 129
  ligneous, 16, 23
  mucopurulent, 117
  papillary, 22, 123
  pseudomembranous, 115, 116, 118
  purulent, 129
  staphylococcal, 22
  swimming pool, 129
  vernal, 22, 23, 34
  verrucose, 119
  viral, 22
Contusion, 168, 188
Copper, 14, 17, 156, 157, 170, 171, 172
Corectopia, 79, 81
Cornea
  abrasions, 167
  amyloid, 11
  arcus senilis, 8, 9
  blood staining, 48
  cornea guttata, 10, 191, 192
  cornea plana, 186
  Descemet's scroll, 167
  descemetocele, 24, 121
  dystrophy
    Fuchs epithelial-endothelial, 10, 191, 192
    granular, 144
    hereditary crystalline stromal, 14
    lattice, 11, 14, 144
    macular 9, 14, 144
    Schnyder, 11
  edema, 167, 191, 192
  Fleischer's ring, 14
  grafts, 39, 40, 43, 169, 188, 191
  healing, 167
  Hudson-Stähli line, 14
  immune ring, 36, 37, 127
  keratinization, 162
  keratitis (*see* Keratitis)
  keratoconus, 14, 87, 192, 194
  keratomalacia, 162
  keratomycosis, 127, 137
  keratopathy (*see* Keratopathy)
  kerectasia, 24
  megalocornea, 77
  mesodermal dysgenesis, 77
  microcornea, 77, 186
  opacification, 134, 155, 163
  perforation, 121
  posterior embryotoxon, 80
  scar, 24
  staphyloma, 24
  ulcer, 24, 31, 116, 120, 125, 127, 132, 178
    amebic, 132
    chronic, 24
    dendritic, 120
    hypopyon, 17
    marginal, 24, 31
    Mooren, 24
    mycotic, 127
    ring, 24, 31
    serpiginous, 24, 116
  vascularization, 40, 52, 53, 67, 154
  Wessely ring, 36
  wounds, 167
Corynebacteria, 118, 162
Corynebacterium xerosis, 162
Cotton-wool spots, 42, 62, 63, 66, 67, 138, 144, 146, 179
Cretinism, 211
Cryosurgery, 177
Cryptococcosis, 127
Cryptophthalmos, 83, 85
Cryptorchism, 79
Crystals, 11, 15, 56, 57, 60, 105, 214
  calcium oxalate, 11
  cholesterol, 11, 56, 57, 60, 214
  cystine, 11
  multiple myeloma, 11, 105
  tyrosine, 11
  urate, 11, 15
Cyanolabe, 159
Cyclitis, 16
Cyclodialysis, 167, 170
  internal, 170
Cyclopia (cyclops), 68, 87
Cylindroma, 100
Cysts, 20, 69, 73, 77, 99, 110, 111, 112, 130, 135, 166, 218
  ciliary body, 111, 112
  colobomatous, 69
  cutaneous, 110
  dermoid, 20, 73, 111, 218
  epidermoid, 20
  epidermoid implantation, 111, 218
  hydatid, 135
  implantation, 111, 166
  intraepithelial, 110
  iris, 69
  Meibomian, 99
  parasitic, 218

pars plana, 111
pilar, 110
pseudocysts, 110
retention, 110
sclera, 77
sebaceous, 110
socket, 111
sudoriferous, 110
toxoplasma "true," 131
Cysticercus cellulosae, 135
Cystinosis, 11, 140
Cytoid bodies, 62
Cytomegalic inclusion disease, 119, 137, 207

Dacryoadenitis, 16, 125, 218
Dacryocystitis, 16, 114, 116, 119
Dalen-Fuchs nodules, 25, 39
Degeneration
    elastotic, 192
    tapetoretinal, 202
Delayed hypersensitivity, 34, 39, 115
Dengue fever, 119
Dermatitis
    allergic, 34
    atopic, 195
    contact, 34
    herpetiformis, 195
    spectacle, 35
Dermatomycoses, 127
Dermatomyositis, 42
Dermoid, 72
Dermolipoma, 72
Descemet's scroll, 167
Descemetocele, 24, 121
Detachment
    ciliochoroidal, 198
    macula, serous, 69
    neuroepithelium, disciform, 219
    retina (see Retina, detachment)
Deuteranope, 144, 159
Diabetes insipidus, 31
Diabetes mellitus, 4, 10, 48, 51, 57, 58, 114, 138, 144-148, 211
Diabetic iridopathy, 146
Diabetic retinopathy, 62, 138, 144, 145
Dichromat, 159
Digitalis intoxication, 164
Diktyoma, 96
Dinitrophenol, 211
Diphtheria, 22, 118
Diphyllobothrium, 136
Diseases (see also syndromes and individual conditions)
    Addison, 163
    bacterial, 115-118
    Batten, 6, 141
    Behçet, 24, 29, 207

Bernheimer-Seitelberger, 140
Best, 144, 159
Bloch-Sulzberger, 143
Bourneville, 86, 159
celiac, 161
Coats, 56, 57, 67
collagen, 62
cretinism, 211
Crouzon, 85, 86
dominantly inherited drusen of Bruch's membrane, 144
Doyne's honeycomb choroiditis, 144, 159
epidemic parotiditis, 119
extramammary Paget, 90
Fabry, 141, 155, 156, 164, 211
familial blepharophimosis, 86
familial dwarfism with stiff joints, 143
familial dysautonomia, 143
familial high density lipoprotein deficiency, 141
familial hyperbeta and prebetalipoproteinemia, 142
familial hyperbetalipoproteinemia, 142
familial hypercholesterolemia, 4
familial hyperchylomicronemia, 142
familial hyperchylomicronemia with hyperprebetalipoproteinemia, 142
Farber's lipogranulomatosis, 141, 155
foot and mouth, 119
fungal, 127, 128, 137
Gaucher, 141
generalized gangliosidosis, 140, 155
globoid cell leukodystrophy, 141
glycolipid lipidosis, 141, 155, 156
Gm_1-gangliosidosis (type I), 140
Gm_1-gangliosidosis (type II), 141
Gm_2-gangliosidosis (type I), 140, 155
Gm_2-gangliosidosis (type II), 140, 155
Gm_2-gangliosidosis (type III), 140
gout, 11
Graves, 162, 164
Hand-Schüller-Christian, 30, 31
hemoglobin-C, 140
hemoglobin-S, 140
hepatolenticular degeneration, 14, 142, 156, 157
hereditary aniridia, 144
hereditary benign intraepithelial dyskeratosis, 3, 15
hereditary color blindness, 144
Hodgkin, 104
Hooft, 143
hydatid, 135
hyperlipoproteinemia (types I-type VI), 142
incontinentia pigmenti, 143
infantile amaurotic family idiocy, 140
Jansky-Bielschowsky, 141

Diseases—continued
  juvenile Gm$_1$-gangliosidosis, 141
  juvenile Gm$_2$-gangliosidosis, 140
  Krabbe, 141
  Kufs, 141
  Letterer-Siwe, 30, 31
  lipomucopolysaccharidosis, 155
  macroglobulinemia, 7
  mandibulofacial dysostosis, 72, 85, 86
  Marie-Strümpell, 24
  metachromatic leukodystrophy, 141, 155
  metaphysial dysostosis (Jansen type), 86
  mucolipidosis I, 155
  mucopolysaccharidoses, 9, 139
  mucoviscidosis, 161
  myotonic dystrophy (myotonia dystrophica), 143, 211
  neuronal ceroid lipofuscinosis, 6, 141, 155, 207
  Newcastle disease of fowls, 22
  Niemann-Pick, 141, 155, 207
  Niemann-Pick, chronic, 155
  Norman-Landing, 140
  Norrie, 143, 159
  oculodentodigital dysplasia, 86, 186
  oculovertebral-auricular dysplasia, 72
  Oguchi, 144
  osteogenesis imperfecta, 77, 86
  Paget disease of bone, 12, 143, 198
  parasitic, 130–136
  Pelizaeus-Merzbacher, 143, 207
  phytanic acid storage disease, 141, 156
  pulseless disease (see Syndrome, aortic arch)
  recessively inherited foveal dystrophy (see Disease, Stargardt)
  Refsum, 141, 156, 207
  rickettsial, 126, 127
  Sandhoff-Jatzhewitz-Pitz, 140, 155
  Sanfilippo, 139, 149, 207
  Sickle cell thalassemia, 140
  Stargardt, 144, 159
  Sticker's progressive arthrophthalmopathy, 143
  Still, 24, 41
  Takayasu, 67
  Tangier, 141
  Tay-Sachs, 140, 155
  vascular, 62
  vitelliform dystrophy of the fovea (see Disease, Best)
  von Hippel-Lindau, 70, 71, 86, 143, 158
  Von Recklinghausen's disease of nerves (neurofibromatosis), 7, 70, 86, 89, 102, 103, 112, 157
  Waldenström, 7
  Wilson, 14, 142, 156, 157
  Wolman, 141

Disorders
  aminoacid metabolism, 149–154
  endocrine, 162, 163
  metabolic, 138–165
  mucopolysaccharides, 149
  protein metabolism, 149–154
Distichiasis, 83
Drugs and toxins, 163–164
Drusen, 1, 12, 19, 82, 83, 201, 219
  optic nerve head, 12, 82, 83
Dulcitol, 149
Dyskeratosis, 3
Dyslipoproteinemia, 141
Dysplasia, 3, 90, 99
Dysproteinemia, 111
Dystrophy, 3, 9, 10, 11, 14, 144, 191, 192
  areolar choroidoretinal, 144
  central choroidoretinal, 144
  central crystalline corneal, 11
  Fuchs epithelial-endothelial corneal, 10, 191, 192
  granular corneal, 144
  hereditary crystalline stromal, 14
  lattice corneal, 11, 14, 144
  macular corneal, 9, 14, 144
  Schnyder's, 11

Ectopia lentis, 81, 152, 156, 157, 186, 212
Ectropion, uveae, 20, 21, 183
Edema, 31, 44, 52–56, 66, 67, 82, 103, 105, 119, 126, 132, 162, 167, 174, 183, 189, 191, 192
  conjunctiva, 56
  cornea, 167, 191, 192
  macula, 174
  optic disc, 31, 44, 52–56, 66, 67, 82, 103, 105, 119, 126, 132, 183, 189
  retina, 56
Elastic fibers, 12
Elastosis
  actinic 12, 89, 193
  solar, 12
Electromagnetic radiation, 172
Elschnig pearls, 211
Emboli
  fat, 61, 179
  fibrin, 61
  infected, 61
  platelet, 61
Embolism, 60, 61, 67, 179
Embryotoxon, posterior, 80
Encephalocele, 73, 158
Endophthalmitis, 16, 37, 39, 133, 134
  nematode, 56, 133
  phacoanaphylactic, 37, 39
  phacotoxic, 39
Entropion, 20, 28, 217
  cicatrical, 28

eyelid, 28, 217
uveae, 20
Eosinophilia, 133
Ephelides, 7
Epicanthic fold, 83, 87
Epidermolysis bullosa, 195
Epinephrine, 59, 163, 164
Episcleritis, 16, 25, 119
Epithelial downgrowth, 166, 188
Epithelial ingrowth (see Epithelial downgrowth)
Epithelioma, benign, 96
Ergot, 211
Erysipelas, 115
Erythema multiforme, 28, 42, 195
Erythema nodosum, 195
Erythrolabe, 159
Eserine, 22
Ethambutol, 163
Exciting eye, 38
Exophthalmos, 29, 31, 67, 102, 158, 162, 163, 164, 218,
  causes, 218
Exudates, 52, 56–58, 138
  hard waxy, 58
  purulent, 26
  retinal, 57, 58
  soft, 62
Eyelid
  ablepharon, 83
  adenoma, 90
  ankyloblepharon, 83
  basal cell carcinoma, 90, 98, 99
  basal cell papilloma, 98
  blepharitis, 16, 115
  blepharospasm, 217
  burns, 28
  café-au-lait spots, 158
  chalazion, 26, 27, 99
  coloboma, 73, 83
  dermatitis (see Dermatitis)
  developmental anomalies, 83, 84
  ectropion, senile, 217
  entropion, 28, 217
  entropion, cicatrical, 28
  epicanthic fold, 83, 87
  erysipelas, 115
  folliculitis, acute, 26
  lacerations, 28
  melanoma, 108
  neoplasms, 90, 97–99
  nevi, 74–76
  ptosis, 30, 132
  scars, 28
  seborrheic keratosis, 90, 98
  symblepharon, 23, 42
  trichiasis, 28, 42
  wart, 119

xanthelasma, 4, 5

Ferry line, 14
Fibrinoid material, 10
Fibroma, 91
Fibromatosis, meningeal, 102
Fibrosarcoma, 91, 176
Fibrous downgrowth, 188
Fibrous dysplasia, 91
Fibrous xanthoma, 91
Filarioidea, 134
Fistula
  carotid-cavernous, 45, 51, 52, 188, 218
  ocular, 170
Fleischer's ring, 14
Fleurette, 93
Flexner-Wintersteiner rosettes, 93
Flies
  black, 134
  blowflies, 136
  botflies, 136
  houseflies, 135
  mangrove, 135
  warble, 136
Floaters, 214
Floccules of Busacca, 18
Fluorescein angiography, 44, 50, 154
Fluorescein solutions, 117
Folliculitis, acute, 26
Fractures, 169
  blow out, 169
Francisella tularensis, 117
Freckles, 7, 74
  uvea, 74
Fusarium, 127

Galactokinase deficiency, 140, 148
Galactose-1-phosphate uridyl-transferase deficiency, 140, 148
Galactosemia, 140, 148, 149, 164, 211
Gangrene, 8
Gasterophilidae, 136
Giant follicular lymphosarcoma, 104
Glaucoma, 14, 29, 31, 43, 48, 52, 59, 64, 65, 69, 79, 81, 89, 92, 95, 107, 123, 144, 154, 155, 157, 158, 164, 168, 170, 181–189, 190, 199, 211, 212, 216
  absolute, 185
  acute, 52, 59, 186, 190, 212
  angle closure, 188
  chronic, 182, 183, 184
  chronic simple, 187
  closed angle, 64, 182, 183, 186
  congenital, 123, 144, 153, 158, 184, 186, 190
  due to
    α-chymotrypsin, 189
    steroids, 189
  hundred day, 64

Index

Glaucoma—*continued*
  low tension, 189
  narrow angle, 186
  open angle, 182, 187, 213, 219
  phacolytic, 188, 211
  phacomorphic, 188
  primary angle closure, 144, 186
  pupillary block, 188
  secondary, 92, 168, 182, 187
  soft, 189
  thrombotic, 64
Glioma
  optic nerve, 6, 9, 89, 91, 101, 102, 112, 157, 158
  retina, 159
Glioneuroma, 90
Gliosis
  intraretinal, 26
  postretinal, 26
  preretinal, 26
Gonococcus, 22, 116
Gonorrhea, 116
Gram-positive cocci, 115-116
Granulation tissue, 19, 167
Granulocytic sarcoma, 105
Granuloma, 116, 133, 188, 218
  eosinophilic, 30, 31
  foreign body, 218
  midline lethal, 30
  Wegener, 30, 31
Granuloma pyogenicum, 19
Granulomatous reaction, 136

Hamartoma, 70, 73
Hartmannella, 132
Hemangioblastoma, 158
Hemangioendothelioma, 71, 91
  benign, 71, 91
  infantile, 71
  malignant, 91
Hemangioma, 70, 71, 72, 84, 108, 218
  angioblastic, 71, 158
  capillary, 70, 71
  cavernous, 71
Hemangiopericytoma, 91
Hematoma, retrobulbar, 218
Hemoflagellates, 131
Hemoglobinopathies, 140, 154
Hemophilus ducreyi, 117
Hemophilus influenzae, 117
Hemorrhage, 31, 47, 48, 49, 105, 117, 121, 138, 144, 154, 166, 168, 169, 178, 179, 187, 188, 212
  anterior chamber, 31, 48, 49, 121, 188, 212
  choroid, 47, 48
  conjunctiva, 47, 178
  expulsive, 48
  macula, 198
  orbit, 169, 218
  retina, 47, 105, 117, 138, 154, 179
  subhyaloid, 48
  vitreous, 48, 117, 144, 157
Hemosiderin, 47, 71
Heterochromia iridis, 31, 151, 199
Heterografts, 33
Histiocytosis X, 30-31, 218
Histoplasma capsulatum, 128
Histoplasmosis, 137
Hole
  macular, 168
  retinal, 206, 207
Homocystinuria, 140, 152, 153, 164, 186, 212
Homograft, 34
Hordeolum
  external, 26
  internal, 26
Horseshoe kidney, 79
Hudson-Stähli line, 14
Humoral immune responses, 33, 35-37, 114
Humoral immunity, 114
Hutchinson's melanotic freckle, 108
Hyaline, 10
Hyalinization, 20
Hyaloid artery, persistent remnants of, 78
Hydrocephalus, 218
Hydroxychloroquine, 163
Hypercalcemia, 12, 13, 30, 156
Hyperemia, 58-59
Hyperplasia, 3
  atypical, 3
  pseudoepitheliomatous, 19
  retinal pigment epithelium, 108
Hypersensitivity, 10, 33
Hypertelorism, 85
Hypertension, 10, 58, 66, 144
Hypertension, malignant, 10, 48, 52, 56, 62, 66
Hyperthyroidism, 162-163, 218
Hypertrophy, 2, 3
Hyphema, 31, 48, 49, 121, 188, 212
Hypocalcemia, 156, 211
Hypoderma bovis, 136
Hypogammaglobulinemia, 114
Hypophosphatasia, 156
Hypoplasia, iris, 78, 87, 186
Hypopyon, 17, 24, 29
  sterile, 17
  ulcer, 17
Hypospadias, 79
Hypotony (hypotonia), 52, 56, 167, 189

Iminophenoxine, 207
Immune ring, 36, 37, 127
Immunoglobulin, 33, 35, 42, 114

Index 229

Immunoglobulin A (IgA), 114
Immunoglobulin E (IgE), 33
Immunoglobulin G (IgG), 35
Immunoglobulin M (IgM), 35
Immunosuppression, 114, 121, 131
Impetigo, 115
Inclusion blennorrhea, 129
Infarction, 8
  hemorrhagic, 154
  retina, 65
Inflammation, 16–32, 107
  granulomatous, 24
  pseudotumor, inflammatory, 28–29, 31, 32
Influenza, 24, 119
Injuries
  alkali, 177, 178, 179
  blast, 169
  chemical, 177–178
  contact lenses, 167
  contrecoup, 168
  chest, 61, 179
  electric, 176
  flames, 176
  neck, 61
  ocular, 166
  physical and chemical, 166–179
  scalds, 176
  strangulation, 179
  thermal, 176–177
  ultrasonic, 177
  visible light, 175
Insudate, 52, 58
Interferon, 114
Intrascleral nerve loop, 77
Inverted follicular keratosis, 90
Inverted folliculoma, 90
Iodoacetate, 207
Iridocyclitis, 16, 17, 25, 29, 43, 117, 119, 123, 126
  acute, 126
  anaphylactic, 43
  chronic, 25, 123
  nongranulomatous, 25
Iridodialysis, 167
Iridodonesis, 212
Iris
  aniridia, 78, 79, 87, 186
  Brushfield spots, 87
  corectopia, 79
  cyst, 69
  diabetic iridopathy, 146
  essential atrophy, 199
  floccules of Busacca, 18
  glycogen storage, 146, 147
  heterochromia iridis, 31, 151, 199
  hypoplasia, 78, 87

  infarcts, 154
  inversion, 167
  iridodialysis, 167
  iridodonesis, 212
  iris bombé, 81, 187, 188
  iritis, 16, 29, 126, 188
  melanoma, 105, 107
  neovascularization, 160
  nevus, 76
  pseudopolycoria, 79
  recession, 167
  rubeosis iridis, 20, 48, 51, 64, 67, 102, 146, 188, 189
  rupture of sphincter pupillae, 167
  tear, 167
  tumors, 105, 111, 113
Iris bombé, 81, 187, 188
Iron 14, 171, 172, 179
Irradiation, 12, 89
  solar, 89
Ischemia
  carotid, 51
  retina, 42, 62, 67, 179
Ischemic optic neuropathy, 67
Isoantigens, 39
Isoniazid, 163

Juvenile xanthogranuloma, 31, 32

Kaposi sarcoma, 91
Kayser-Fleischer ring, 14, 156, 157
Keratic precipitates, 18, 24
  conglomerate, 18
  lardaceous, 18
  mutton fat, 18
  waxy, 18
Keratitis, 16, 28, 119, 121, 126, 130, 137
  disciform, 121
  herpes, 121, 137
  interstitial, 23
  mycotic, 137
  sclerosing, 16
  superficial, 23
  verrucose, 119
Keratoacanthoma, 90, 97, 99, 112
Keratoconjunctivitis, 22, 34, 118, 119, 122, 125, 128, 196
  acute actinic, 175
  cicatrizing, 128
  epidemic 22, 119, 122
  phlyctenular, 34, 115
  sicca, 41, 42, 196
Keratoconus, 14, 87, 192, 194
Keratomalacia, 162
Keratomycosis, 127, 137
Keratopathy
  band, 12, 13, 30, 41, 162

230 Index

Keratopathy—continued
　bullous, 182, 191
　chronic actinic, 14
　exposure, 28, 117, 196, 217, 219
　lipid, 154
　neuropathic, 117, 196
　punctate, 120, 123
Keratosis
　actinic, 99
　solar, 99
Kerectasia, 24
Klebsiella pneumoniae, 117
Koeppe nodules, 18
Koplik spots, 125
Krukenberg spindle, 183

Lacerations, 28, 166, 167, 168
　retina, 168
Lacrimal gland, 16, 29, 84, 85, 90, 99–101, 125
　adenoid cystic carcinoma, 90, 100, 101
　dacryoadenitis, 16, 125, 218
　ectopic lacrimal gland, 84, 85
　mixed tumor, 90, 99–100
　mucoepidermoid carcinoma, 90, 101
　neoplasms, 90, 99–101
　oncocytoma, 90
　pleomorphic ademona, 99
　sarcoidosis, 29
Lacrimal sac, 16, 90, 101, 110, 116, 119
　adenocarcinoma, 101
　dacryocystitis, 16, 114, 116, 119
　hydrops, 110
　mucocele, 110
　neoplasms, 101
　papilloma, 90, 101
　squamous cell carcinoma, 101
Lagophthalmos, 217
Lasers, 173, 175, 176, 179
Lecithin-cholesterol acetyltransferase (LCAT) deficiency, 142
Leiomyoma, 90, 91, 101
Leishmania, 131
Leishmania tropica, 131
Leishmaniasis, 131
Lens
　aphakia, 70
　cataract (see Cataract)
　dislocation, 188, 212
　displaced, 188, 212
　ectopic, 81, 152, 156, 157, 186, 212
　Elschnig pearls, 211
　glaukomflecken, 183
　lenticonus, 79
　Mittendorf dot, 81
　opacities, 156
　spherophakia, 78, 153, 186, 212
　subluxation, 80, 212
Lens capsule
　exfoliation, 212
　pseudoexfoliation, 212, 219, 220
　warts, 79
Lenticonus, posterior, 79
Lentigines, 7
　senile, 7
Lentigo-maligna melanoma, 108
Leprosy, 20, 23, 116–117
　lepromatous, 117
　tuberculoid, 117
Leukemia, 48, 104, 105, 111, 128
　granulocytic, 105
　lymphatic, 105
　myeloid, 105
Leukocoria, 81, 93
Lipemia retinalis, 146, 155
Lipid proteinosis, 143
Lipidoses, 4, 155
Lipofuscin, 6
Lipoma, 91
Liposarcoma, 91
Liver spots, 7
Loa loa, 134
Lymphangiectasia, 72
Lymphangioma, 72, 218
Lymphoepithelial lesion, benign, 43
Lymphogranuloma venereum, 115, 130
Lymphoma, 90, 103–104, 114, 121
　Burkitt, 104, 119
　diffuse lymphocytic, 104
　diffuse undifferentiated, 104
　follicular, 104
　histiocytic, 104
　malignant, 91, 112
　nodular, 104
Lysozyme, 114

Macroglobulin, 111
Macroglobulinemia, 7, 48
Macular degeneration, senile, 204
　disciform, 204, 205
Macular star, 57
Maculopathy, 119, 179, 204
　cystoid, 204
　photic, 179
　subacute sclerosing panencephalitis, 119
Malta fever, 117
Mastigophora, 131
Measles, 123, 125, 207
　German, 119, 123
　red, 119, 125
Medulloepithelioma, 15, 90, 95, 96
　embryonal, 95
　teratoid, 90, 96
Megalocornea, 77

Meibomian gland
  adenocarcinoma, 90, 99
  adenoma, 90
  cyst, 99
Melanocytoma, 75
Melanocytosis, oculodermal, 76
Melanomas, 38, 69, 71, 77, 90, 105-109, 113, 176, 189, 190
  choroid, 106
  conjunctiva, 109
  cutaneous, 109
  epithelioid, 107
  eyelid, 108
  fascicular, 107
  intraepithelial, 109
  iris, 105, 107
  malignant, 90, 109, 112, 113
  mixed, 107
  necrotic, 107
  primary malignant, 112
  ring, 107
  spindle A, 107
  spindle B, 107
  uvea, 77, 105, 107, 112, 113
Melanosis
  acquired conjunctival, 7, 90, 108
  dermal, 75
  episcleral, 76
  ocular, 76, 77
  scleral, 76
Mellaril, 164
Membrane
  cyclitic, 21
  iridovitreal, 20
  preretinal, 20
  pupillary, 20, 81
  retrocorneal, 20
  retrovitreal, 20
  vitreocorneal, 20
Meningeal fibromatosis, 102
Meningioma, 55, 91, 102, 103, 157, 218
Meningocele, 73
Meningoencephalocele, 218
Mer-29, 164, 211
Mesodermal dysgenesis of the cornea, 77
Metaplasia, 1
Methanol, 164
Methoxyflurane, 11
Microaneurysms, 48-50, 58, 64, 66, 67, 105, 138, 146, 154
Microcornea, 77, 186
Microfilaria, 134, 135
Microorganisms, 17, 19, 33, 114
Microphthalmia (microphthalmos), 79, 123, 131
Microphthalmos with orbital cysts, 69, 70
Migraine, 60, 61

Milia, 110
Miotics, 211
Mittendorf-dot, 81
Mixed tumors, 90, 99, 100
  malignant, 100
Mollusca fibrosa, 158
Molluscum contagiosum, 119, 123, 124
Mongolian spot, 75
  extrasacral, 75
Mongolism (see Syndrome, Down)
Morgagnian corpuscles, 210
Mucoceles, 110, 218
Mucormycosis, 127, 128, 148, 218
Multiple myeloma, 7, 11, 48, 111, 112
Mumps, 125
Muscidae, 136
Mycobacterium, 34, 115, 116
Mycobacterium fortuitum, 116
Mycobacterium leprae, 116, 117
Mycobacterium tuberculosis, 115, 116
Myiasis, 136
Myleran, 211
Myopia, 147, 152, 157, 211, 219
  progressive, 219
  simple, 219
  stationary, 219
Myositis, 218
Myxedema, 162
Myxoma, 91

Naegleria, 132
Naphthalene, cataracts due to, 211
Necrosis, 8, 93, 107, 129
  caseation, 8
  coagulation, 8
  fibrinoid, 66, 133
  gummatous, 8
  liquefaction, 8
Necrotic tissue, 176
Neisseria gonorrhoeae, 114, 116
Nematodes, 56, 132
Neoplasia, 3-4
Neoplasms, 89-113
  metastatic, 71, 109
Neovascularization, 19, 67, 154, 160
  iris, 160
  retina, 50, 67, 146
Neurilemoma, 103
Neuroblastoma, 92, 109
Neurofibroma, 89, 90, 91, 103, 157, 158
Neurofibrosarcoma, 91, 103
Nevus, 74-76, 108, 113
  blue, 75, 76
  cellular blue, 76
  choroid, 75
  conjunctiva, 75
  dermal, 74

Index

Nevus—*continued*
 iris, 76
 junctional, 7, 74, 109
 of Ota, 76
 subepithelial, 74
 uvea, 74, 113
Nevoxanthoendothelioma, 31
Nicotinic acid, 162
Nodular fasciitis, 19, 32, 218
North American blastomycosis, 127
Nuclear explosions, 173
Nyctalopia, 162

Ochronosis, 12, 149
Oestridae, 136
Onchocerca volvulus, 134
Onchocerciasis, 207, 218
Oncocytoma, 90
Ophthalmia neonatorum, 22
Ophthalmia nodosa, 136
Opsin, 162
Optic disc and nerve
 Bergmeister papilla, 82
 drusen, optic nerve head, 12, 83
 glaucomatous cupping, 183, 184
 glioma, 6, 9, 89, 91, 101, 102, 112, 157, 158
 meningioma, 91, 102
 tumors, 112
Optic atrophy, 55, 66, 116, 126, 132, 164, 182, 215, 216
 ascending, 215
 cavernous, 9, 184, 216
 descending, 215
 Schnabel, 184
Optic neuritis, 16, 47, 54, 119, 163
Optic neuropathy, ischemic, 67
Optic pits, 69, 87, 88
Orbit
 abscess, 218
 anomalous, 175, 218
 cellulitis, 16, 107, 218
 cysts, 111
 ectopic brain tissue, 218
 emphysema, 169, 218
 encephalocele, 73
 fractures, 169
 meningocele, 73
 meningoencephalocele, 218
 nodular fasciitis, 218
 phycomycosis, 127
 pseudotumors, 28, 29, 31, 32
 tumors, 90, 91
 xanthomatosis of orbital fat, 218
Ossification, 1, 2
Osteogenic sarcoma, 89, 91, 95
Osteoma, 91
Oxalosis, 14

Oxycephaly, 85
Oxygen toxicity, 160–161, 164

Pannus
 glaucomatous, 183
 trachomatous, 129
Panophthalmitis, 16, 116, 118, 125, 127
Papilledema, 31, 44, 52–56, 66, 67, 82, 103, 105, 119, 126, 132, 162, 183, 189
Papillitis, 16
Papilloma, 90, 99, 101, 121
 lacrimal sac, 90, 101
 squamous, 90, 99
Papillomatosis, 3, 99
 recurrent juvenile conjunctival, 99
Paragonimus westermani, 136
Parakeratosis, 3
Pasteurella tularensis, 117
Pemphigus, 195
Pemphigus vulgaris, 195
Periarteritis nodosa, 10, 24, 41
Peripheral anterior synechiae, 20, 64, 187, 188
Pernicious anemia, 62
Persistent hyperplastic primary vitreous, 80, 81, 87
Peter's malformation (anomaly), 77, 88, 186
Phagocytosis, 19
Phakoma, 159
Phakomatosis, 70, 87
Pharyngoconjunctival fever, 119, 122
Phenothiazine, 207, 211
Phlyctenule, 35
Phthisis bulbi, 1, 2, 107, 159, 198
Phycomycosis, 127
Pinguecula, 12, 14, 192–194
Plagiocephaly, 85
Plasma cell myeloma, 91
Pleomorphic adenoma, 99
Plexiform neurofibroma, 158
Pneumococcus, 116
Poliosis, 42
Polyarteritis nodosa (*See* Periarteritis nodosa)
Polycythemia, 22, 52, 56
Posterior synechiae, 20
Precipitates, 18
 keratic, 18, 24,
 preretinal, 18
 retinal, 18
Presbyopia, 209
Progressive systemic sclerosis, 42
Proptosis, 30, 84, 91, 92, 128, 217, 219
Prostatitis, 24
Protanope (red blind), 144, 159
Protozoa, 130–132
Pseudocysts, 130, 208

Index 233

Pseudoglaucoma, 189
Pseudohermaphroditism, 79
Pseudomembrane, 22, 122
Pseudomonas aeruginosa, 117, 137
Pseudopolycoria, 79
Pseudopterygium, 24
Pseudorosettes, 93
Pseudosarcomatous fasciitis, 19
Pseudotumors, 26, 28, 31, 32
   orbit, 28, 29, 31, 32
   retina, 26
Pseudoxanthoma elasticum, 12, 143, 198
Psittacosis, 128
Psoriasis, 195
Pterygium, 14, 193, 195
Ptosis, 30, 132
Pupils
   Argyll Robertson, 126
   eccentric, 199
Pyridoxine deficiency, 163

Quinine, 164
Quinone, 150

Radiant energy, 172
Radiation, 112, 172-175
   infrared, 172, 173, 174
   ionizing, 174, 175
   solar, 173
   ultraviolet, 173-175
Recession of iris, 167
Red eye, 59
Regional enteritis, 24
Retina
   angioma, 57
   degeneration, 9, 171, 202, 203, 204, 207, 219
      cobblestone, 204
      lattice, 204
      peripheral cystoid, 9, 153, 202, 203, 219
      reticular cystoid, 203
      typical cystoid, 203
   detachment, 11, 51, 68, 91, 107, 144, 157, 170, 175, 179, 203, 206-208, 209, 211, 219, 220
   dysplasia, 82, 87
   infarction, 65
   ischemia, 42, 62, 67, 179
   lacerations, 168
   neovascularization, 50, 51, 67, 146, 160
   occlusovascular disease, 62
   phakoma, 159
   precipitates, 18
   pseudocysts, 208
   pseudotumor, 26
   retinitis, 137
   retinopathy (*see* Retinopathy)

   retinoschisis, 203-204, 220
   tear, 168, 169, 175, 208
   venous disease, 63-64
Retinene, 162
Retinitis, 137
   cytomegalic inclusion, 137
Retinitis pigmentosa, 144, 206, 211
Retinitis proliferans, 50, 144, 154, 207
Retinoblastoma, 12, 48, 51, 56, 71, 81, 89, 90, 92-95, 96, 97, 107, 113, 133
   diffuse infiltrating, 93
   endophytic, 93
   exophytic, 93
Retinol, 161
Retinopathy, 138, 144-146, 174
   diabetic, 62, 138, 144-146
   eclipse, 174
   exudative diabetic, 138
   hypertensive, 42, 60, 61, 67
   pigmentary, 155, 156, 205, 206, 207
   of prematurity, 160-161, 207
   proliferative, 87
Retinoschisis, 203-204, 220
   juvenile, 144, 203-204
   senile, 203
Retrobulbar neuritis, 16, 126
Retrolental fibroplasia, 160
Rhabdomyosarcoma, 91, 101, 112, 113
Rheumatic fever, 10
Rheumatoid arthritis, 10, 16, 24, 41
   juvenile, 41
Rheumatoid factor, 41
   nodule, 10, 41
Rhinosporidium seeberi, 127
Rhodopsin, 161, 162, 201
Ribose, 149
Rickettsia prowazekii, 127
Rickettsia rickettsii, 127
Rieger anomaly, 80, 85, 86, 186
Ringschwiele, 208, 209
River blindness, 134
Rocky Mountain spotted fever, 126
Rosacea, 195
Rosenthal fibers, 6, 102
Roseola, 126
Rosettes, 93, 95
Roundworms, 132-135
Rubella, 78, 86, 119, 123, 137, 186, 207, 211, 212
Rubeola, 115, 123, 125
Rubeosis iridis, 20, 48, 51, 64, 67, 102, 146, 188, 189

S-thalassemia, 140
Sandfly fever, 119
Sarcoidosis, 20, 24, 29, 30, 32, 156, 218
Sarcoma, 89, 91

Index

Sarcoma—*continued*
  reticulum cell, 104, 112
Sattler's veil, 167
Scaphocephaly, 85
Schistosoma, 136
Schwalbe ring, 80
Schwannoma, 89, 90, 91, 103, 157
Sclera
  alkaptonuria, 149
  blue, 77
  calcification, 13
  cyst, 77
  episcleritis, 16, 25, 119
  intrascleral nerve loop, 77
  melanosis, 76
  pigmentation, 149, 150
  quinone, 149
Scleritis, 16, 41, 43, 119, 218
  annular, 16, 41
  brawny, 41
  necrogranulomatous, 43
  posterior, 16, 218
Sclerocornea, 77
Scleroderma, 42, 211
Scleromalacia perforans, 41
Seborrhea, 27
Seborrheic keratosis, 7, 90, 98
Seborrheic wart, 98
Shigella, 117
Shwartzman phenomenon, 10
Sickle cell disease, 48, 60, 140, 154, 198
Sickle cell hemoglobin C disease, 140
Sickle cell thalassemia, 140
Siderosis bulbi, 171, 207
Silver, 14
Silver nitrate, 17
Silver wiring, 66
Simulium, 134
Smallpox, 119, 125, 207
Snow blindness, 175
Sodium fluoride, 207
Soft chancre, 117
South American trypanosomiasis, 131
Sparganosis, 135
Sparganum, 136
Spherophakia, 78, 153, 186, 212
Sphingolipidoses, 140, 164
Spirochetes, 126
Sporotrichosis, 127
Spring catarrh, 34
Sprue, 161
Staphylococcus, 22, 28, 34, 115, 116, 191
Staphylococcus aureus, 115
Staphyloma, 24, 166, 183, 185
  ciliary, 183, 185
  corneal, 24
  equatorial, 183
  intercalary, 183
Stenosis
  nasolacrimal duct, 84
  vascular, 60
Steroids, 114, 164, 189, 211
Stocker line, 14
Streptococci, 115, 116
Streptococcus viridans, 116
Striated muscle, 1, 96
Sty, 115
Subacute sclerosing panencephalitis, 119, 125, 137
Sulfatide lipidosis, 141
Suppuration, 17
Symblepharon, 23, 42
Sympathetic ophthalamia, 20, 24, 37, 38, 39, 43, 170
Sympathetic ophthalmitis, 38, 39
Sympathizing eye, 38
Synchisis scintillans, 214
Syndromes, 85–87
  Alport, 86, 143
  anterior segment cleavage, 80, 85
  aortic arch, 48, 67
  Axenfeld, 80, 186
  Bardet-Biedl, 143, 207
  Bassen-Kornzweig, 141, 207
  Battered baby 56
  Behçet, 24, 29, 207
  Brusa-Torricelli-Miller, 79
  Carpenter, 86
  Chédiak-Higashi, 143, 150
  Cockayne, 207
  coloboma of iris-anal atresia-extra chromosome, 86
  Cornelia de Lange, 86, 143
  craniofacial dysostosis, 86
  craniosynostosis, 85
  Crouzon, 85, 86
  Down, 83, 84, 86, 87, 211, 212
  Edwards, 86, 87
  Ehlers-Danlos, 83, 143, 198
  Fliessinger-Leroy, 29
  Foster Kennedy, 55, 103
  Franceschetti-Treacher Collins, 86
  Fraser, 86
  Goldenhar, 72, 86
  Hallerman-Streiff, 86
  Hallervorden-Spatz, 139
  Hermansky-Pudlak, 139
  hidrotic ectodermal dysplasia (Marshall type), 143
  Horner, 199
  Hunter, 139, 149, 207
  Hurler, 139, 207
  Kearn-Sayre, 207
  kinky hair, 142

Laurence-Moon, 143, 207
Leri's pleonosteosis, 86
Lignac-Fanconi, 140
Lowe's oculocerebrorenal, 86, 140, 153, 186
Maffucci, 71
Manoteaux-Lamy, 139, 149
Marchesani, 186
Marfan, 143, 156, 157, 186, 188, 212
Marinesco-Sjögren, 143
Menke, 142
Mieten, 86
Mikulicz, 30
milk-alkali, 156
Morquio, 139, 149
Parinaud oculoglandular, 30, 115
Patau, 86, 87
Pierre Robin, 86, 186
poikiloderma congenita, 143
Reiter, 25, 29, 31
Riley, 86
Riley-Day, 143
Romberg, 199
Rothmund-Thomson, 143
Rubinstein-Taybi, 86
Scheie, 139, 149, 207
Schwartz, 86
sea blue histiocyte, 155
Sjögren, 41, 43
Smith-Lemli-Opitz, 86
Spielmeyer-Sjögren-Batten, 141
Stevens-Johnson, 22, 23, 42, 195
Sturge-Weber, 70, 71, 86, 186
Terson, 48
Urbach-Wiethe, 143
Usher, 207
Vogt-Koyanagi-Harada, 24
Waardenburg-Klein, 86, 199
Weill-Marchesani, 78, 86, 212
Werner, 86
Wyburn-Mason, 84, 86
Syneresis, 214
Synophthalmia, 68
Synostosis, 85
Syphilis, 8, 20, 23, 60, 115, 126, 130, 207
    congenital, 23
Systemic lupus erythematosus, 10, 42

Taenia echinococcus, 135
Taenia saginata, 135
Taenia solium, 135
Tapeworms, 135–136
Tear
    iris, 167
    retina, 168, 169, 175, 208
Teratoid medulloepithelioma, 90, 96
Teratoma, 70, 74, 218

Thalassemia, 83
Thalidomide, 69, 86
Thalium, 211
Thioridazine, 164
Thrombi, mural, 60
Thromboangiitis obliterans, 60
Thrombosis
    cavernous sinus, 218
    central retinal vein, 105
Toxocara canis, 132, 137
Toxocariasis, 132, 133
Toxoplasma, 114, 130, 131
Toxoplasma gondii, 114, 130
Toxoplasmosis, 20, 24, 32, 130, 131, 132, 137, 207
Trachoma, 10, 22, 23, 28, 123, 128, 129, 130, 137
Transudates, 52–56
Trauma, 166–170
Trematodes, 136
Treponema pallidum, 114, 115, 126
Treponema pertenue, 126
Treponematoses, 126
Trichiasis, 28, 42
Trichinella, 133
Trichinella spiralis, 133, 134
Trichoepithelioma, 90
Triparanol, 164
Trypanosoma, 131
Trypanosoma cruzi, 131
Trypanosomiasis, African, 132
Tsutsugamushi fever, 126
Tuberculosis, 8, 20, 23, 24, 114, 116, 121
    cutaneous, 116
    meningeal, 116
Tuberous sclerosis, 12, 70, 86, 143, 159
Tularemia, 115, 117
    oculoglandular, 117
Tyndall phenomenon, 17
Typhoid, 207
Typhus
    epidemic, louse-borne, 127
    Old World tick, 127
    tropical, 126
Tyrosine, 11
Tyrosinemia, 139, 150

Ulcerative colitis, 24, 25
Urticaria, 34
Uvea
    ectropion uveae, 20, 21, 183
    entropion uveae, 20
    melanoma, 105–107, 112
    nevus, 74–75, 113
Uveitis, 16, 25, 32, 36, 38, 41, 43, 119, 125
    anterior, 16, 24, 25
    granulomatous, 25

Index

Viruses—continued
  heterochromic, 25
  immunologic, 35
  lens-induced, 43
  nongranulomatous, 24
  phacotoxic, 39
  posterior, 16, 24
  recurrent, 25

Vaccination, 120
Vaccinia, 119, 137
Varicella, 121
Variola, 119, 125
Vasculitis, allergic, 42
Vasoconstriction, 61
Vasodilation, 58, 59
Verruca plana, 119, 121
Verruca vulgaris, 119, 121
Viruses, 82, 89, 104, 118-125
  adenoviruses, 22, 115, 119, 122, 137
  arboviruses, 119
  coxsackie, 119
  cytomegalovirus, 3, 86, 114, 119, 121, 131
  dengue fever, 119
  enteroviruses, 119
  Epstein-Barr, 104, 119
  herpes simplex, 24, 118, 119, 137
  herpes simplex virus type I, 118
  herpes simplex virus type II, 114, 118
  herpes virus varicellae, 119, 121
  herpes viruses, 118, 119, 120, 121
  herpes zoster, 24, 119, 121, 122, 137
  human wart virus, 119
  mumps, 119
  Newcastle disease virus, 119, 125
  papovaviruses, 119, 121
  picornaviruses, 119
  poxviruses, 119
  rhinovirus of cattle, 119
  rubeola virus, 119
  sandfly fever virus, 119
  togaviruses, 119
  varicella zoster, 119
  yellow fever virus, 119
Visceral larva migrans, 133
Visual purple (see Rhodopsin)
Vitamin A, 161, 162
Vitamin D, 14, 156, 162
Vitamin deficiencies and excesses, 161-162
Vitiligo, 38
Vitreous, 81
  abscess, 26, 133
  amyloid, 11
  asteroid hyalitis, 213, 214, 215
  floaters, 214
  hemorrhage, 48
  opacities, 11, 24, 213
  persistent hyperplastic primary, 80, 81, 87
  syneresis, 214

Wart, 3, 98, 119
  senile, 98
Wegener's granuloma, 31
Wegener's granulomatosis, 24, 29, 30, 218
Wessely ring, 36
Wilms tumor, 79, 87, 95
Wounds
  healing, 187
  perforating, 166, 169, 170
  surgical, 169-170

Xanthelasma, 4, 5
Xanthoma, 4, 155
Xanthomatosis of orbital fat, 218
Xenon photocoagulators, 175
Xeroderma pigmentosum, 89, 143, 195
Xerostomia, 41
X-rays, 175
Xylose, 149

Yaws, 126
Yellow fever, 119